NANOBIOTECHNOLOGY

Concepts and Applications in
Health, Agriculture, and Environment

NANOBIOTECHNOLOGY

Concepts and Applications in Health, Agriculture, and Environment

Edited by

Rajesh Singh Tomar, PhD

Anurag Jyoti, PhD

Shuchi Kaushik, PhD

AAP | APPLE ACADEMIC PRESS

Apple Academic Press Inc.
4164 Lakeshore Road
Burlington ON L7L 1A4, Canada

Apple Academic Press Inc.
1265 Goldenrod Circle NE
Palm Bay, Florida 32905, USA

Exclusive worldwide distribution by CRC Press, a member of Taylor & Francis Group

International Standard Book Number-13: 978-1-77188-824-0 (Hardcover)
International Standard Book Number-13: 978-0-42929-275-0 (eBook)

Library and Archives Canada Cataloguing in Publication

Title: Nanobiotechnology : concepts and applications in health, agriculture, and environment / edited by Rajesh Singh Tomar, PhD, Anurag Jyoti, PhD, Shuchi Kaushik, PhD.

Other titles: Nanobiotechnology (Burlington, Ont.)

Names: Tomar, Rajesh Singh, editor. | Jyoti, Anurag, editor. | Kaushik, Shuchi, editor.

Description: Includes bibliographical references and index.

Identifiers: Canadiana (print) 20190240857 | Canadiana (ebook) 20190240865 | ISBN 9781771888240 (hardcover) | ISBN 9780429292750 (ebook)

Subjects: LCSH: Nanobiotechnology.

Classification: LCC TP248.25.N35 N36 2020 | DDC 660.6—dc23

Library of Congress Cataloging-in-Publication Data

Names: Tomar, Rajesh Singh, editor. | Jyoti, Anurag, editor. | Kaushik, Shuchi, editor.

Title: Nanobiotechnology : concepts and applications in health, agriculture, and environment / edited by Rajesh Singh Tomar, Anurag Jyoti, Shuchi Kaushik.

Other titles: Nanobiotechnology (Tomar)

Description: Burlington, ON ; Palm Bay, Florida : Apple Academic Press, [2020] | Includes bibliographical references and index. | Summary: "This new book, Nanobiotechnology: Concepts and Applications in Health, Agriculture, and Environment, presents a broad conceptual overview regarding the synthesis, applications, and toxicological aspects of nanobiotechnology. It focuses on the entrance into and interaction of nanomaterials in the human body, which has generated intense scientific curiosity, attracting much attention as well as increasing concern from the nanomaterial-based industries and academia throughout the world. This book looks at the scientific aspects of nanomaterials used in many applications of biosciences, taking an interdisciplinary approach that encompasses medicine, biology, pharmacy, physics, chemistry, engineering, nanotechnology, and materials science. The volume covers the basics of nanosciences and nanotechnology; different schemes and routes of synthesis; and various biological applications, including sensing, medicine, drug delivery systems, and remediation. Further, special chapters are devoted to nanotoxicology and the developing risk factors associated with nanosized particles during use along with the ethical issues related to nanobiotechnology. The expert group of authors was chosen for their distinguished expertise and belong to the academic and industrial worlds, creating a broad perspective. This will be valuable resource for materials scientists, chemists, and biologists who want to explore the captivating world of nanobiotechnology"-- Provided by publisher.

Identifiers: LCCN 2019056253 (print) | LCCN 2019056254 (ebook) | ISBN 9781771888240 (hardcover) | ISBN 9780429292750 (ebook)

Subjects: MESH: Nanotechnology | Biotechnology

Classification: LCC R857.N34 (print) | LCC R857.N34 (ebook) | NLM QT 36.5 | DDC 610.28--dc23

LC record available at https://lccn.loc.gov/2019056253
LC ebook record available at https://lccn.loc.gov/2019056254

Apple Academic Press also publishes its books in a variety of electronic formats. Some content that appears in print may not be available in electronic format. For information about Apple Academic Press products, visit our website at **www.appleacademicpress.com** and the CRC Press website at **www.crcpress.com**

About the Editors

Rajesh Singh Tomar, PhD, MSc, MPhil

Director, Amity Institute of Biotechnology,
Dean (Academics), Amity University (Madhya Pradesh),
Gwalior, India

Rajesh Singh Tomar, PhD, MSc, MPhil, is the Professor and Head/Director of the Amity Institute of Biotechnology, Dean (Life Sciences) and Dean (Academics), Amity University Madhya Pradesh, Gwalior, India. He is the Founder Campus Head of Amity, Gwalior. Professor Tomar has more than 27 years of teaching and research experience in various disciplines of life sciences. Prof. Tomar is actively engaged in teaching, research, and innovation. His research interest includes nanobiotechnology and environmental biotechnology. He has supervised three PhD scholars for award of PhD, and currently eight scholars are pursuing their PhD under his supervision. Presently, he is the PI of a research project funded by the Madhya Pradesh Council of Science and Technology, Bhopal, India. He has to his credit more than 110 research papers published in peer-reviewed national and international journals of repute. He has also published four books and authored a dozen book chapters published by national and international publishers. He has to his credit seven patents, of which three have been published. He has also delivered lectures in more than 35 national and international seminars, symposia, conferences and workshops. He has been conferred as a fellow member of Chemical, Biological & Environmental Engineering Society, Hong Kong. He is member of various international and national scientific bodies and reviewer and editor of national and international journals.

Anurag Jyoti, PhD

Assistant Professor, Amity Institute of Biotechnology, Amity University (Madhya Pradesh), Gwalior, India

Anurag Jyoti, PhD, is an Assistant Professor at the Amity Institute of Biotechnology, Amity University Madhya Pradesh, Gwalior, India. He has obtained a master's degree in Biotechnology from the prestigious Indian Institute of Technology Roorkee in 2006 and PhD in Biotechnology from CSIR–Indian Institute of Toxicology Research, Lucknow, in 2012. He has qualified joint CSIR-UGC JRF-NET in the year 2005. He has seven years of teaching and research experience and has 35 research papers to his credit. He has co-supervised one PhD scholar for the award of PhD. Presently, he is the PI of a research project funded by MPCST, Bhopal, India. He has also authored two books and five book chapters. He has presented papers at various national and international conferences. Dr. Anurag is actively engaged in teaching, research, and innovation. His research interests include nanobiotechnology and environmental microbiology. Dr. Anurag has won a Young Scientist Award of the Madhya Pradesh Council of Science and Technology, Bhopal, India, in 2013. He is in the editorial board and reviewer of various reputed journals and has is life member of various international and national scientific bodies.

Shuchi Kaushik, PhD

Ex-Assistant Professor, M.P. Forensic Science Laboratory, Madhya Pradesh, India

Shuchi Kaushik, PhD, is Scientific Officer, Madhya Pradesh Forensic Science Laboratory, M.P., India. She has more than twelve years of experience in teaching and research. She has participated in more than thirty national and international conferences/seminars, presenting her research work. She is actively engaged in research activities and is a member of reputed scientific societies and journals. Dr. Kaushik has published 36 research papers in journals of national and international repute. She is co-supervising two PhD scholars. She has published three book chapters and two books in the form of lab manuals. She has won a Young Scientist Award of the Madhya Pradesh Council of Science and Technology, Bhopal in 2014. Her research interests include microbiology and biotechnology.

Contents

Contributors .... *ix*

Abbreviations .... *xiii*

Preface .... *xvii*

Introduction .... *xix*

1. **Nanobiotechnology: An Ocean of Opportunities** **1**
   Mala Trivedi, Rachana Singh, Parul Johri, and Rajesh K. Tiwari
2. **Nanomaterials: Novel Preparation Routes, Characterizations, and Applications** **23**
   Juhi Saxena and Anupam Jyoti
3. **Nano-Biomedicine: A Next-Generation Tool for Effective and Safe Therapy** **35**
   Vikas Shrivastava, Pallavi Singh Chauhan, and Rajesh Singh Tomar
4. **Nanotheranostics: Next Generation Diagnosis and Therapy for Cancer** **45**
   Sharmistha Banerjee, Shuchi Kaushik, and Rajesh Singh Tomar
5. **Recent Advancement of Nanobiotechnology in Cancer Treatment** **57**
   Abhinav Shrivastava
6. **Role of Nanoparticles in Common Cereals** **71**
   Raghvendra Kumar Mishra, Arunchand Rayaroth, and Rajesh Singh Tomar
7. **Nanotoxicology: A Threat to Human Health and Environment** **87**
   Sushmita Shrivastava and Rajesh Singh Tomar
8. **Impact of Nanotoxicity on Human Health: An Overview** **93**
   Bharti Prakash and S. Mathur
9. **Nanomaterials: Properties, Synthesis, Characterizations, and Toxicities** **111**
   Mohit Agarwal, Rajesh Singh Tomar, R. K. Singh, Anil Kumar Meena, and Anurag Jyoti

10. Nanoparticles: Method of Production and Future Prospective 125

Neha Sharma and D. S. Rathore

11. Nanoinformatics: An Emerging Trend in Cancer Therapeutics 135

Medha Pandya, Snehal Jani, Vishakha Dave, and Rakesh Rawal

12. Thin Films: An Overview 163

Pankaj Kumar Mishra

13. Pharmaceuticals Polymers in Medicine 183

Divya Singh and Narendra Pal Singh Chauhan

14. Utilizing Nanotechnology for Environmental Cleaning 199

Kuldip Dwivedi, Shivani Yadav, and V. K. Yadav

15. Implementation of Nanotechnology in Agriculture System: A Current Perspective 211

Raghvendra Saxena, Manish Kumar, and Rajesh Singh Tomar

Index.. 229

Contributors

Mohit Agarwal
Amity Institute of Biotechnology, Amity University (Madhya Pradesh), Maharajpura, Gwalior – 474 005, India

Sharmistha Banerjee
Amity Institute of Biotechnology, Amity University (Madhya Pradesh), Gwalior – 474005, India, E-mail: sbanerjee@gwa.amity.edu

Narendra Pal Singh Chauhan
Department of Chemistry, Bhupal Nobel's University, Udaipur, Rajasthan, India, E-mail: narendrapalsingh14@gmail.com

Pallavi Singh Chauhan
Amity Institute of Biotechnology, Amity University (Madhya Pradesh), Gwalior – 474005, India

Vishakha Dave
Research Scholar, Department of Physics, M.K Bhavnagar University, Bhavnagar, India, Tel.: +91-9408728323, E-mail: Vishakhadave1612@gmail.com

Kuldip Dwivedi
Department of Environmental Science, Amity University (Madhya Pradesh), Gwalior – 474005, Madhya Pradesh, India, E-mail: dwivedikul2012@gmail.com

Snehal Jani
Assistant Professor, Amity School of Applied Sciences, Amity University (Madhya Pradesh), Gwalior, India, Tel.: +91-8866014107, E-mails: sneh.jani@gmail.com; scjani@gwa.amity.edu

Parul Johri
Amity Institute of Biotechnology, Amity University Uttar Pradesh, Lucknow Campus, Near Mallhour Railway Station, Mallhour, Gomtinagar Extension, Lucknow – 28, Uttar Pradesh, India

Anupam Jyoti
Amity Institute of Biotechnology, Amity University Rajasthan, Jaipur, India, E-mail: ajyoti@jpr.amity.edu, anupamjyoti@rediffmail.com

Anurag Jyoti
Amity Institute of Biotechnology, Amity University (Madhya Pradesh), Maharajpura, Gwalior – 474 005, India, E-mail: ajyoti@gwa.amity.edu

Shuchi Kaushik
M.P. Forensic Science Laboratory, Madhya Pradesh, India, E-mail: skaushik@gwa.amity.edu

Manish Kumar
Amity Institute of Biotechnology, Amity University (Madhya Pradesh), Gwalior, Madhya Pradesh, India

S. Mathur
Department of Zoology, SPC Government College, Ajmer, Rajasthan, India

Anil Kumar Meena
Division of Toxicology, CSIR-Central Drug Research Institute, Jankipuram Extension, Lucknow – 22 6031, India

Pankaj Kumar Mishra
Amity School of Applied Sciences, Amity University (Madhya Pradesh), Gwalior, India, E-mail: pmishra@gwa.amity.edu

Raghvendra Kumar Mishra
Amity Institute of Biotechnology, Amity University (Madhya Pradesh), Gwalior, India, E-mail: rkmishra@gwa.amity.edu

Medha Pandya
Assistant Professor, The K.P.E.S Science Collage, M.K Bhavnagar University, Bhavnagar, India, Tel.: +91-9662031001, E-mail: megsp85@gmail.com

Bharti Prakash
Department of Zoology, SPC Government College, Ajmer, Rajasthan, India, E-mail: dr.bharti.prakash@gmail.com

D. S. Rathore
Department of Biotechnology, Government Kamla Raja Girls P. G. (Autonomous) College, Gwalior, Madhya Pradesh, India

Rakesh Rawal
Professor, Department of Life Sciences, Gujarat University, Ahmadabad, India, Tel.: +91-9925244855, E-mail: rakeshmrawal@gmail.com

Arunchand Rayaroth
Amity Institute of Biotechnology, Amity University (Madhya Pradesh), Gwalior, India, E-mail: arunrayaroth@gmail.com

Juhi Saxena
Dr. B. Lal Institute of Biotechnology, Malviya Industrial Area Jaipur, India

Raghvendra Saxena
Amity Institute of Biotechnology, Amity University (Madhya Pradesh), Gwalior, Madhya Pradesh, India, E-mail: rsaxena@gwa.amity.edu

Neha Sharma
Amity Institute of Biotechnology, Amity University Madhya Pradesh, Gwalior, India, E-mail: drneha16may@gmail.com

Abhinav Shrivastava
Department of Biotechnology, College of Life Sciences, Cancer Hospital and Research Institute, Gwalior, Madhya Pradesh, India, E-mail: abhi.shri76@gmail.com

Sushmita Shrivastava
Amity Institute of Biotechnology, Amity University Madhya Pradesh, Gwalior, India E-mail: rssush@gmail.com

Vikas Shrivastava
Amity Institute of Biotechnology, Amity University (Madhya Pradesh), Gwalior – 474005, India, E-mail: vshrivastava@gwa.amity.edu

Divya Singh
Department of Chemistry, Amity School of Engineering and Technology,
Amity University (Madhya Pradesh), Gwalior, India, E-mail: drdsingh18@gmail.com

R. K Singh
Division of Toxicology, CSIR-Central Drug Research Institute, Jankipuram Extension,
Lucknow – 22 6031, India

Rachana Singh
Amity Institute of Biotechnology, Amity University Uttar Pradesh, Lucknow Campus,
Near Mallhour Railway Station, Mallhour, Gomtinagar Extension, Lucknow – 28,
Uttar Pradesh, India

Rajesh K. Tiwari
Amity Institute of Biotechnology, Amity University Uttar Pradesh, Lucknow Campus,
Near Mallhour Railway Station, Mallhour, Gomtinagar Extension, Lucknow – 28,
Uttar Pradesh, India, E-mail: rktiwari@lko.amity.edu

Rajesh Singh Tomar
Amity Institute of Biotechnology, Amity University (Madhya Pradesh), Maharajpura,
Gwalior – 474005, India, E-mail: rstomar@amity.edu

Mala Trivedi
Amity Institute of Biotechnology, Amity University Uttar Pradesh, Lucknow Campus,
Near Mallhour Railway Station, Mallhour, Gomtinagar Extension, Lucknow – 28,
Uttar Pradesh, India

Shivani Yadav
Department of Life Sciences, ITM University, Gwalior, Madhya Pradesh, India

V. K. Yadav
Division of Crop Improvement, IGFRI, Jhansi, Uttar Pradesh, India

Abbreviations

Ag	silver
AgNPs	silver nanoparticles
AMF	alternating magnetic field
ATP	adenosine triphosphate
AuNPs	gold nanoparticles
BBB	blood-brain barrier
BFO	basic formal ontology
CaPN	calcium phosphate nanoparticle
CAS RN	chemical abstracts service registry number
CASP	critical assessment of protein structure prediction
CeO_2	cerium dioxide
CNTs	carbon nanotubes
CSCNT	cup-stacked CNT
CTCs	circulating tumor cells
CuO	copper oxide
CVD	chemical vapor deposition
DACH	diaminocyclohexane
DLA%	damaged leaf area percentage
DLS	dynamic light scattering
DNA	deoxyribonucleic acid
EDX	energy dispersive x-ray analysis
ENMs	engineered nanomaterials
EPR	enhanced permeation and retention
FDA	Food and Drug Administration
FISH	fluorescent *in situ* hybridization
FITC	fluorescein isothiocyanate
FTIR	Fourier transform infrared
FWHM	full width at half maximum
GHS	globally harmonized system
GI	gastrointestinal
GRAS	generally recognized as safe
GSH	glutathione
HIV	human immunodeficiency virus

HT	high tension
IONPs	iron oxide nanoparticles
ISA	investigation study assay
LSPR	localized surface plasmon resonance
LT	low tension
MBD	methyl-CpG-binding domain
MD	molecular dynamics
MgO NPs	magnesium oxide nanoparticles
MLL	mixed-lineage leukemia
MMT	montmorillonite
MPS	mononuclear phagocyte system
MRI	magnetic resonance imaging
MSNs	mesoporous silica nanoparticles
MWCNTs	multi-walled carbon nanotubes
$NaBH_4$	sodium boro-hydride
NCI	National Cancer Institute
NCp7	nucleocapsid protein 7
NMs	nano-materials
NPO	nanoparticle ontology
NPs	nanoparticles
NSPs	nanoscale particles
NZVI	nanosized zero-valent ion
OCNR	Office of Cancer Nanotechnology Research
OECD	Organization for Economic Cooperation and Development
OFPs	oncogenic fusion proteins
OWL	ontology web language
PCB	polychlorinated biphenyl
PDT	photodynamic therapy
PEG	polyethylene glycol
PLGA	poly(lactic-coglycolic acid)
PSMA	prostate-specific membrane antigen
PTX	paclitaxel
PVD	physical vapor deposition
PVP	polyvinyl pyrrolidone
QDs	quantum dots
ROG	radius of gyration
SC704	single cross 704
SEM	scanning electron microscopy

SNPs	silicon nanoparticles
SPR	surface plasmon resonance
SWCNTs	single-walled carbon nanotubes
TAB	tab-delimited
TE	tissue engineering
TEM	transmission electron microscopy
Tf	Transferrin
TiO_2 NPs	TiO_2 nanoparticles
TiO_2	titanium dioxide
TNF	tumor necrosis factor
XPS	x-ray photoemission spectroscopy
ZnO	zinc oxide
ZVI	zero-valent iron

Preface

Nanotechnology is a fast-evolving discipline that already produces outstanding basic knowledge and industrial applications for the benefit of our society. The first applications of nanotechnology appear mainly in the field of materials science; applications in the field of life sciences are still developing. The combination of nanosciences and biotechnology shows promises for the development of new and innovative tools for applications in health and environmental sciences.

The definitions and overall view of nanobiotechnology have been covered in the first chapter by Trivedi et al. The authors have described the basics of nanobiotechnology and overall applications in various fields. Biosynthesis of nanoparticles using bacteria, yeast, fungi, and plants has been discussed by Saxena et al. in Chapter 2. They have further described the different characterization methods and overview of potential applications of nanomaterials in different fields. In Chapter 3, Shrivastava et al. have presented the novel applications of nanomaterials in medicine. Nano-biomedicine has the potential to serve as a next-generation tool for an effective and safe therapy. Potential applications of nanobiotechnology in cancer detection and therapy have been discussed in Chapters 4 and 5 by Banerjee et al., and Shrivastava et al., respectively. Mishra et al. have described the role of nanotechnology in cereals in Chapter 6.

No doubt that nanotechnology has enormous scope and potential for the betterment of society, but still, their harmful aspects need to be noticeable, which affects each and every trophic level of the ecosystem. Shrivastava et al. and Prakash et al. have presented the harmful effects nanomaterials on human health and environment in Chapters 7 and 8, respectively. Chapter 9 emphasizes the overall synthesis, properties, characterization, and toxicity of nanomaterials. Agarwal et al. have presented a detailed overview of cytotoxicity, genotoxicity, organ toxicity, and an estimate of how much of a substance causes a kind of harm. In Chapter 10, Sharma and Rathore have presented a comprehensive overview of methods of nanoparticle synthesis and their future prospective. Nanoinformatics is the new area of nanotechnology, in which the designing and technological development of nanoparticle production can be performed. It is the interface of nanotechnology, biotechnology, and bioinformatics. In Chapter 11, Pandya et al. have covered a wider perspective on nanoinformatics and its applications in cancer therapeutics. In Chapter 12, Dr. Mishra has covered thin films and given an overview of their synthesis.

Introduction

Nanobiotechnology is the recent most technology in the present era. It focuses on the manipulation of materials at the nano-size level, followed by a combination with biomolecules. The possibility to exploit the structures and processes of biomolecules for novel functional materials, biosensors, and medical applications has created the rapidly growing field of nanobiotechnology. With the rapid development of nanotechnology and its applications, a wide variety of nanostructured materials are now used in commodities, pharmaceutics, cosmetics, biomedical products, and industries. While nanoscale materials possess more novel and unique physicochemical properties than bulk materials, they also have an unpredictable impact on human health. The entrance into and interaction of nanomaterials in the human body have generated intense scientific curiosity, attracting much attention as well as increasing concern from the nanomaterial-based industries and academia across the world.

The proposed book contains information about different pillars of nanobiotechnology that will assist scientists and students in learning the fundamentals and cutting-edge nature of this new and emerging science. Focusing on materials and building blocks for nanotechnology, leading scientists from around the world will share their knowledge and expertise. The book presents an extraordinary and thorough overview of the emerging field of nanobiotechnology for biologists, material scientists, chemists and, and others from many diverse fields.

This book will be a valuable reference source for academicians, scientists, research professionals working in R&D laboratories, students, college, and university professors working in the field of nanobiotechnology. The book is intended for a broad audience working in the fields of biotechnology, materials science, environmental science, medicine, and toxicology.

CHAPTER 1

Nanobiotechnology: An Ocean of Opportunities

MALA TRIVEDI, RACHANA SINGH, PARUL JOHRI, and RAJESH K. TIWARI*

Amity Institute of Biotechnology, Amity University Uttar Pradesh, Lucknow Campus, Near Mallhour Railway Station, Mallhour, Gomtinagar Extension, Lucknow – 28, Uttar Pradesh, India

**Corresponding author. E-mail: rktiwari@lko.amity.edu*

ABSTRACT

Nanobiotechnology is an amalgamation of engineering and molecular biology, leading to the development of devices with specificity and high sensitivity. This branch has developed to a great extent; its major applications are in the field of health and medicine, besides that in some or other ways in energy and environment. Applications of nanoparticles in diagnosis, gene therapy, drug delivery, pathogen detection, and tissue engineering (TE) are discussed in detail. An in-depth study on recent developments in the area of health and diagnostics using nano-objects (quantum dots (QDs), nanoparticles, nanotubes, etc.) is discussed. Future applications of the field in the treatment of important diseases in the form of smart drugs and the development of new devices are also explored. Beyond biotechnology, nanomaterials find their use in various other fields too, including defense services, electronics, etc. With so many applications and future prospects, there are some risk factors as well as challenges too in the form of toxicity and regulations. The objective of this chapter is to describe the potential benefits and impacts of nanobiotechnology in various areas and its future prospects.

1.1 INTRODUCTION

The prefix "nano" derives from the Greek word for dwarf. One nanometer is equal to one-billionth of a meter. Or about the width of 6 carbon atoms or 10 water molecules [73].

The field of nanotechnology took form in the early 1980s, when the development of scanning probe microscopy allowed researchers, for the first time, to observe and manipulate individual atoms. Through years of research and advances, manipulations of nanoscale objects have become less cumbersome, and the field is now making rapid advances. Nanomaterials are more than just tiny versions of bulk material. Nanoscale materials possess different physical and chemical properties than larger-scale matter because their size is sufficiently small that quantum mechanics dictates some of their properties. For example, gold is normally solid, but at the nanoscale, it becomes a liquid at room temperature. Another key property of nanomaterials is that as a particle gets smaller, it's relative surface area increases, and its electronic structure changes. The unique properties of nanomaterials due to quantum effects and surface-area effects allow new applications for many materials. The ability to create and design novel materials is, ultimately, why nanotechnology promises to make a huge impact on so many fields.

Nanobiotechnology represents the convergence of nanotechnology and biotechnology, yielding materials and products that use biological molecules in their construction or are designed to affect biological systems. Several applications of nanobiotechnology include:

- Engineering biomolecules from non-biological use, such as DNA-based computer circuits.
- Using nanotechnology tools such as medical diagnostic devices and medical imaging to study biology.
- Combining nanomaterials with biological systems for outcomes such as targeted drug therapies.

Nanomaterials can be constructed by a top-down approach or a bottom-up approach. In the top-down approach, the nano-builder reduces the size of a material of interest until it reaches nanoscale proportions. In the bottom-up approach, nanostructures are built atom-by-atom or molecule-by-molecule by either manipulating individual atoms or creating conditions where components can self-assemble. Self-assembly is much faster than building nanomaterials atom by atom; engineering new nanomaterials that self-assemble may provide a critical manufacturing solution. The field of biology provides many examples of self-assembling molecules, such as proteins that fold into a specific configuration.

Novel nanomaterials are being developed for biotechnology and medical applications. The combination of nanotechnology and biotechnology has led to the emerging field of nanobiotechnology. Research in nanotechnology, biotechnology, and information technology transects progressively. Multidisciplinary research involving medical researchers, biologists, chemists, physicists, materials scientists, and engineers is boosting nanobiotechnology forward. The union of these once-distinct fields is the essence of nanobiotechnology. Because nanomaterials are of the same scale as biological molecules, nanomaterials may open new possibilities for monitoring and intervening in biological systems.

Nanomaterial and devices provide unique opportunities to advance medicine is referred to as "nanomedicine" and could impact diagnosis, monitoring, and treatment of diseases as well as control and understanding of biological systems [52].

Researchers are using nanobiotechnology to create new contrast agents for cell imaging, to deliver gene therapy, and to analyze cellular processes. Nanomaterials are being investigated as a potential platform for cell transplantation in the treatment of diseases such as Parkinson's and diabetes. Nanotechnology may also play an important role in developing tissue-engineering materials that are more compatible with the body. On-site diagnostic work to assess the presence or activity of a particular molecule can be faster and more sensitive using nanoscale tags [61].

Analyses of signaling pathways by nanobiotechnology techniques might provide new insights into disease processes, thus identifying more efficient biomarkers and understanding the mechanisms of action of drugs [29]. Advances in the manipulation of the nanomaterials permit the binding different biomolecules, such as bacteria, toxins, proteins, and nucleic acids [14]. Nanotechnology is relatively new, and although the full scope of contributions to these technological advances in the field of human health care remains unexplored, recent advances suggest that nanobiotechnology will have a profound impact on disease prevention, diagnosis, and treatment [46].

Nanotechnology is the technology involved in engineering products at the atom or molecular level by reducing their size to an extent affecting its properties [23]. It has applications in every sphere of life, including cosmetics, clothing, sporting goods, personal care, sunscreen, medical, and tissue engineering (TE). Therefore, when applications are in the above-mentioned fields, it is called nanobiotechnology. The cell size of living organisms is approximately 10 μm, and cell organelles are much smaller, less than a micron. The building block of cells, proteins have a size of about 5 nm, which is comparable with the dimensions of the smallest man-made nanoparticles (NPs) [56].

Science of nanobiotechnology has various applications, including: fluorescent biological labels [8, 12, 69], drug, and gene delivery [38, 50], bio-detection of pathogens [17], detection of proteins [47], probing of DNA structure [39], TE [37, 16], tumor destruction via heating (hyperthermia) [75], separation, and purification of biological molecules and cells [45] MRI contrast enhancement [70], phagokinetic studies [52]

In an experiment application of carbon nanotubes (CNTs) for transferring gene of interest in plants, *Arabidopsis* and *Glycyrrhiza glabra* was demonstrated [22]. Instead of normal CNTs, cellulose modified cup-stacked CNT (CSCNT—cellulase, lengths between 1μ m–100 μm) used to create nanoholes in the cell wall through which CSCNT with adsorbed biomolecules can move intracellularly, without damaging cell.

Foodborne diseases are very frequent nowadays; three levels of monitoring are required in such cases starting from production to processing and then distribution to point-of-sale [25]. The conventional method of pathogen detection is time taking depending on the type of pathogen, as some will take less than 24 hours, and some more than a weeks' time. Other constraints with the traditional method are pathogen count and environmental conditions, with the change in conditions pathogen, regain virulence, and cause an infection that, in turn, causes health risks [49]. Nanomaterials can easily be manipulated by loading with various biomolecules, for example, nucleic acid, proteins, bacteria, and toxin [34]. Using nanomaterial for the development of biosensors has many advantages, and one of them is in the study of receptors. It does not require the removal of receptors from the lipid membrane of the cell, a compulsory step in other assay methods [25]. There are reports for the use of gold and silver nanoparticles (AgNPs) for the development of biosensors to detect the pathogen in food items [31]. The antimicrobial activity of these metallic salts and complexes (ionic silver) is due to the bonding of metallic ions to biomacromolecular components. Metallic cations bind to negatively nucleic acids and/or proteins leading to change in their structure [1, 9].

In this chapter, we will discuss the use and applications of nanobiotechnology.

1.2 APPLICATIONS OF NANOBIOTECHNOLOGY

1.2.1 *MEDICAL APPLICATIONS*

Two of the most exciting and promising domains of nanotechnology for advancements are health and medicine. Nanotechnology offers potential developments in pharmaceuticals, medical imaging and diagnosis, cancer treatment, implantable materials, tissue regeneration, and multifunctional platforms combining several of these modes of action [20, 61].

1.2.1.1 DIAGNOSIS

One primary goal in nanobiotechnology is the design of new methodologies to diagnose a number of diseases at an early stage with cheaper material and more sophisticated equipment than is possible today [48]. More research is currently being performed in this area. The utilization of metal and semiconductor NPs in biomedical applications has been demonstrated by many research groups. In 2007, Aaron et al. [78] have shown that 25-nm gold nanoparticles (AuNPs), when conjugated with anti-epidermal growth factor receptor monoclonal antibodies, can be efficiently used as *in vivo* targeting agents for imaging cancer markers, specifically epidermal growth factor receptors. The AuNps result in a dramatic increase in signal contrast compared to other antibody-fluorescent dye targeting agents [65]. Nanobodies have the potential to be a new generation of antibody-based therapeutics and to be used in diagnostics for diseases such as cancer. The advantages of nanobodies to developing therapeutics are the extremely stable and bind antigen with nanomolar affinity, a high target specificity, and low toxicity, the ability to combine the advantages of conventional antibodies with important features of small-molecule drugs, and their ability to be produced cost-effectively on a large scale [29]. An option for the use of antibodies in molecular biomedical is aptamers.

These molecules are chemically stable and easily produce single-stranded nucleic acid molecules. One example of an application of aptamers in diagnosis is the work of Niedzwiecki and colleagues [48]. In this work, nanopores and aptamers were combined to detect a single molecule of the nucleocapsid protein 7 (NCp7), a protein biomarker of the HIV-1 virus, with high sensitivity [46, 48].

An interesting tool being developed today to be utilized in tumor diagnosis is RNA NPs [27, 59, 76]. Although several researchers are adverse to RNA nanotechnology, due to the susceptibility of RNA to RNase degradation and serum instability, Shu and colleagues have developed toolkit in which homogeneous RNA NPs were obtained, which targeted cancer exclusively *in vivo* without accumulation in normal organs and tissues. Functionalized NP aggregating fluorescence imaging techniques, known as quantum dots (QDs), have the potential for real-time and non-invasive visualization of biological events *in vivo* [60].

1.2.1.2 GENE THERAPY

Gene therapy is the latest development for the treatment or prevention of genetic disorders by correcting defective genes responsible for disease development based on the delivery of repaired genes or the replacement of incorrect ones [3]. The most common approach for correcting defective genes is the insertion of a normal gene into a nonspecific location within the genome to replace a nonfunctional gene. An abnormal gene could also be swapped for a normal gene through selective reverse mutation, which returns the gene to its normal function [28]. Mammalian cells typically have a diameter of a few microns, and their organelles are within the nanometer range. The use of nano-devices has the advantage of entering the cells more easily when compared to larger devices, and they can, therefore, interact better with the cells or at least in a different way [33]. The use of nanotechnology in gene therapy could be applied to replace the currently used viral vectors with potentially less immunogenic nanosize gene carriers.

Therefore, the delivery of repaired genes or the replacement of incorrect genes is the area in which nanoscale objects could be introduced successfully [54].

1.2.1.3 DRUG DELIVERY

Controlled delivery systems are used to improve the therapeutic efficacy and safety of drugs by delivering them to the site of action at a rate dictated by the need of the physiological environment [66], which in turn would reduce both toxicity and side effects [55]. Electrospun nanofibers may serve as a promising delivery vehicle as a result of their 3D nano-sized features that closely resemble those of the ECM [15]. By this technique it is possible to incorporate biological molecules by using an emulsion or directly in a polymer solution [2, 53].

NPs are another tool of nanotechnology which is under intense investigation for drug delivery. They can principally be fabricated by lipids and polymers [33, 41, 58]. Polymeric compounds that are currently being used in drug products include poly (DL-lactic-coglycolic acid) (PLGA) polyvinyl alcohol, poly (ethylene-co-vinyl acetate), polyimide, and poly (methylmethacrylate) [33]. Co-delivery is an alternative for the administration of different drugs, which by the conventional therapeutic method, cannot be used together. Therefore, nanoscale systems can be used to assist the delivery of incompatible drugs [58].

1.2.1.4 TISSUE ENGINEERING (TE)

The limitations related to the use of allografts and xenografts, as well as increasing life expectancy of the population, have headed scientists around the world to hunt more for alternatives. Therefore, research in this area aims to apply the principles of cell transplantation and engineering to construct biological substitutes. These, in turn, are used in an attempt to restore and maintain the normal function of organs and tissues previously diseased or injured [4, 35]. Thus, the focus of TE is to repair or reconstruct lost or damaged tissue through the use of growth factors, cell therapy, injectable biopolymers, and biomaterials, which serve as support for the development of the cells [67]. Cells interact with the environment around them through thousands of interactions on a nanometric scale. Therefore, the goal of TE on a nanoscale is to create biomaterials that direct interactions between cells and their micro-environment, by the creation of nanoscale molecular signals of biological interest. Thereby, the cells receive, process, and respond to information presented in the surrounding environment; these actions being essential for the control of cell behavior [72]. From the techniques used to construct biomaterials to be cultivated with cells, electrospinning is the most widely studied, and it has also been demonstrated to give the most promising results in terms of TE applications. It is a highly versatile method of transforming solutions, mainly made from polymers, in continuous filaments with diameters ranging from a few micrometers to nanometers. Through this method, the fibers can be obtained randomly or in an ordered way. Electrospinning works by the electrostatic principle, where the solution is supplied to the system via syringe and is subjected to a difference in electrical voltage, yielding solid fiber at the end of the process [26]

The skin is an organ that serves as a physical barrier to the external environment and consists primarily of epidermis and dermis [7, 40]. The skin acts as a protective barrier; any injury caused to it should be repaired quickly and efficiently. In the case of burns and chronic wounds, the treatments available are unsatisfactory to prevent scar and promote healing of the patient. Thus, the regeneration of the skin is an important field for TE [43, 74]. The substitutes fabricated for use in TE normally act as supplementary dermal templates and improve wound healing [77]. Several studies are being conducted in an attempt to create an ideal substitute that could overcome the limitations that comes across so far in the current skin substitutes.

Most of them use polymeric matrices associated with stem cells, and in some cases, biological agents, such as growth factors and plant substances, are also used. Other cells, such as fibroblasts and keratinocytes, are largely used too [13, 30, 44, 63]. In the central nervous system, degeneration of neurons or glial cells or any unfavorable change in the extra-cellular matrix of neural tissue can lead to a wide variety of clinical disorders, such as Alzheimer's, Huntington's, and Parkinson's diseases, as well as traumatic damage such as spinal cord injury. The biggest concern here is that almost all diseases in this system lead to permanent loss of functions [21].

Bone injuries occur for many reasons, including degenerative, surgical, and traumatic processes, resulting in severe pain and disability for millions of people worldwide [24, 36]. Currently, the most widely used synthetic bioactive bone substitute is calcium phosphate-based materials [71].

Because of its limited capacity for spontaneous repair due to its lack of vascularity and poor availability of chondrocytes and progenitor cells, cartilage cannot be restored to its normal function and structure after damage caused by trauma, disease, and accidents. Therefore, TE, as a potential approach to regenerate cartilage tissue, holds good promise. The cells that could be used to treat diseases of cartilage can be either differentiated cells, such as chondrocytes, or undifferentiated cells, such as mesenchymal stem cells [5].

The prime objective of TE is to damage control either by tissue repairing or reconstruction of damaged tissues by prevailing methods. The use of nanotechnology in TE is by creating a biomaterial that promotes interaction between cells and their micro-environment. Now cells receive, process, and respond to information of the micro-environment [72].

1.2.1.5 PATHOGEN DETECTION

There are many outbreaks of food disease reported which clearly demand to monitor food-borne pathogens throughout. The food pathogens may be present in low numbers in a sample for analysis, which makes the detection difficult. Traditional detection methods for pathogen determination like colony count estimation are laborious and time-consuming with completion ranging from 24 h for *E. coli* to 7 days for *Listeria monocytogenes*, and these pose significant difficulties for quality control of semi-perishable foods. Pathogen numbers can also be underestimated using these methods due to microorganisms entering viable but nonculturable states due to environmental stress [49]. Upon restoration from this state by, for example, an increase in temperature of the cells, microorganisms can regain the ability to cause infection, thus posing a health risk. Advances in the manipulation of these nanomaterials permit binding of different biomolecules such as bacteria, toxins, proteins, and nucleic acids [34]. One of the major advantages of using nanomaterials for biosensing is that because of their large surface area, a greater number of biomolecules are allowed to be immobilized, and this consequently increases the number of reaction sites available for interaction with a target species. This property, coupled with excellent electronic and optical properties, facilitates the use of nanomaterials in label-free detection and in the development of biosensors with enhanced sensitivities and improved response times [25].

Biosensors are currently used in the areas of target identification, validation, assay development, lead optimization and absorption, distribution, metabolism, excretion, and toxicity. They are best suited to applications using soluble molecules and overcome many of the limitations that arise with cell-based assays. Biosensors are particularly useful in the study of receptors because they do not require the receptor to be removed from the lipid membrane of the cell, which can be necessary with other assay methods. Single-walled carbon nanotubes (SWCNTs) have been used as a platform for investigating surface-protein and protein-protein binding, as well as to develop highly specific electronic biomolecule detectors. Non-specific binding on nanotubes, a phenomenon found with a wide range of proteins, is overcome by the immobilization of polyethylene oxide chains. A general method is followed that entails the selective recognition and binding of target proteins, conjugating their specific receptors to polyethylene-oxide functionalized nanotubes. These arrays are attractive because no labeling is required, and the entire assay can be done in the solution phase. This, combined with the sensitivity of nanotube electronic devices, provides highly specific electronic sensors for detecting clinically important biomolecules, such as antibodies associated with human autoimmune diseases [7].

QDs are colloidal semiconducting fluorescent NPs consisting of a semiconductor material core (cadmium mixed with selenium or tellurium), which has been coated with an additional semiconductor shell (usually zinc sulfide). Due to their unique size-dependent fluorescence properties and photostability, QDs are widely used in place of traditional fluorescent dyes (fluorescein isothiocyanate: FITC). Functionalized QDs have been used as labels for DNA probing of genomic DNA and in fluorescent *in situ* hybridization (FISH) assays [69].

Metallic NPs, such as gold and silver, have been used for signal amplification in numerous biodiagnostic devices. AuNPs have been used in a variety of optical and electrical assays. The electrical properties of the AuNPs were harnessed for the development of a piezoelectric biosensor, for real-time detection of a food-borne pathogen [31].

Elemental silver and silver salts have been well known as antimicrobial agents in curative and preventive health care for centuries. The antimicrobial activity of the silver salts and complexes (ionic silver) is generally based on the bonding of metallic ions in various biomacromolecular components. Cationic silver targets and binds to negatively charged components of proteins and nucleic acids, thereby causing structural changes and deformations in bacterial cell walls, membranes, and nucleic acids [1]. Silver ions are generally well known to interact with a number of electron donor functional groups like thiols, phosphates, hydroxyls, imidazoles, indoles, and amines. Accordingly, it is believed that silver ions that bind to DNA block transcription while those that bind to cell surface components interrupt bacterial respiration and adenosine triphosphate (ATP) synthesis.

1.2.2 FUTURE PROSPECTUS OF NANOBIOTECHNOLOGY

Predicting the future and the pace of development of any science or technology is difficult. An additional complication is that technology or science can develop in unexpected directions and be useful in ways that no one can envisage. Almost all the current applications of nano are flaccid, and most involve adding a nanomaterial to a regular material as a way of improving its performance. In the present section, we will focus on the future of nanobiotechnology. Majorly, the future of this upcoming science could be listed under the following heads:

1. **Qdots and Smart Drugs for Cancer Treatment:** Most of the applications of nanotechnology in medicine are still under development. Qdots (Quantum dots) that basically target identify and localize the cancerous cells in a body can unravel a great deal of information about the molecular events happening in a cancerous cell and also in the early diagnosis of cancer [32].

   Today the potential side effects of most of the drugs used are frightening. The use of electronically controlled "smart" drugs can minimize these effects. This major advancement in drug delivery and release system is boon of the developments made in nanomaterials (NMs). There are many proposed "smart" medical implants that can release drugs on their requirements when exposed to certain stimuli like ultraviolet rays or electronic impulses. But these current techniques still need to be refined before they are ready for prime time.

2. **Military Applications:** Many "smart weapons" are being proposed for improving and strengthening the defense capabilities against missiles, artillery, and mortars. The rapid development in

the nanotechnology has enabled a significant improvement in the different components of the solid-state laser system that makes laser weapons deployable.

The forthcoming technology where scientists are thinking out of the box to create Harry Potter's invisibility cloak will provide soldiers with the ultimate protection of invisibility during war operations or intelligence and surveillance activities.

3. **Next Generation Computer Processing:** The future quantum computers are at the forefront. These computers will use nanotechnology to shrink their size and increase their processing speed and power as these computers will not be based on the digital 1's and 0's but on a new technology "quantum bits" or "qubits." These qubits will no longer have RAM, ROM, or DRAM but will enjoy MRAM (Magnetoresistive Random Access Memory) [10].

   Now let us switch to DNA computing, which uses the basics of molecular biology and biochemistry for computing processes. DNAzymes and nanobots do not exist so far, but when they do, futurists foresee the possible uses for nanorobots. These include molecular manufacturing (nanofactories) and medical nanobots that push through the bloodstream doing repairs and fighting against infection [18, 19].

4. **Oil Industry:** The oil and automotive industry are also not untouched with the latest advancements made in nanotechnology. Many nanoparticles are used today as fuel additives for reducing toxic emission and fuel consumption. The future is focusing on the development of sensors for temperature, pressure, and stress even under extremely harsh conditions and imaging contrast agents. Nanomaterials are expected to be used as an integrated part of a smart structure composed of various elements, including control devices and actuators. Nanometric thin films and composites with nanostructured fillers are being planned to use for preparing innovative corrosion-resistant material solutions. These can be further used for nanolayered corrosion inhibitors in pipes and tanks. Inspired by the success

of zeolites, nanomembranes are being planned to develop. These nanomembranes will enhance the exploitation of tight gas for removing impurities. Nanoporous and nanoparticular materials are also very gifted to administer the environmental, health, and safety risks resulting from the presence of CO_2 and H_2S in hydrocarbon mixtures [42].

5. **Energy Generation and Utilization:** Today, the current global energy demand is satisfied by fossil fuel coal, crude oil, and natural gas. There is a great need to invest in the field of renewable energy. Nanotechnology can enhance the energy generation aided with cost reduction by the implementation of nano-silicon cells, anti-reflecting glasses in photovoltaic. In the future, nanotechnology can also contribute to the optimization of wind power utilization and inter alia by manufacturing high strength and lightweight materials for rotor blades. Nanotechnology may also chip in the future to the optimization of energetic biomass employment and the development of new conversion methods and efficient utilization of pesticides and fertilizers via nanosensors (Table 1.1).

To add on to the present applications of nanotechnology, many potentially revolutionalized proposals have been put forth for the implementation of nanotechnology in various other fields also, like:

a. Generating clean water for safe drinking.
b. Creation of educational tools (e.g., NanoHub) for teaching science concepts and interaction of probes with the physical world.
c. Creation of nanotools for imaging, measuring, integrating, manipulation, and modeling purposes.
d. Development and application of preemptive and personalized medicines.
e. Implementation of "nanoceuticals," particularly in the food and dairy industry.

TABLE 1.1 Various Applications of Nano-Scale Products

S. No.	Field of Application	Applications
1	Medical field	Drugs and medical devices.
2	Electronics	Batteries, phones, LEDs. OLED, anti-bacterial, and anti-static coating in the keyboard, mouse, cell phones, DVD coating, computer processors and chips.
3	Food industry	Energy drinks, plastic wraps, cutting boards, antibacterial utensils, nano tea, chocolate shakes, canola active oil, nutritional supplements, and food storage containers.
4	Toys and baby goods	Stain resistance plush toys, anti-bacterial baby pacifiers, mugs, milk bottles, anti-bacterial stuff toys, play stations.
5	Personal health care products	Hearing aids, body wash, contact lenses, tooth powder, shampoo, deodorants, insect repellents, bandages, man-made skins, home pregnancy tests.
6	Sports industry	Golf balls and clubs, Tennis rackets and balls, baseball bats, hockey sticks, wet suits, shoe insoles, anti-fogging coatings.
7	Automotive industries	Oil and air filters, car wax, engine oil, tiers, catalysts to improve fuel consumption.
8	Textile industry	Wrinkle-free clothes, stain-resistant apparel, anti-bacterial, and anti-odor clothes, flame retardant fabrics, UV resistant, and protective clothing.
9	Cosmetics industry	Moisturizers, skin creams, cleansers, skin creams, lipstick, mascaras, makeup foundations, makeup removal.
10	Household items	Anti-bacterial Furniture and mattresses, filters, air purifiers, anti-bacterial UV resistant paint, solar cells, cleaning products, disinfectant sprays, fabric softeners.

1.3 CHALLENGES AND ISSUES RELATED TO NANOBIOTECHNOLOGY

Despite several applications of nanomaterials in various fields, there are still some challenges faced by this nascent science. The exact usage and quantity of nonmaterial are not yet known to its full extent. Further, it is difficult to determine how and where theses nanomaterials are used, as many manufacturers do not reveal it. Many products are banned with the word "nano" as an advertisement strategy. Many nanotechnology inventions are not accepted by society.

There are several environmental issues based on the biodegradability of nonmaterial, and if free nanoparticles are being released in air, water, or soil as a pollutant, we don't know yet if certain nanoparticles will correlate themselves as a non-biodegradable pollutant. We are also unaware of the methods through which these particles could be removed from air or water, as most of the traditional filters are not suitable for this type of task. There could be many health hazards of engineered nanoparticles that need to be properly accessed and evaluated during their fabrication, storage, distribution, applications, and dispose [6].

All the above-discussed challenges could be broadly streamlined into four major points:

1. The main challenge is to develop instruments that can access exposure of the engineered NMs into the air, water, and soil;
2. The next challenge is to develop applicable methods to determine and detect the toxicity of NMs in the upcoming decade;
3. Evaluating the exact impact of NMs on human health and environment over the entire life span and proposing models for it is another major challenge in the upcoming field nanobiotechnology; and
4. Developing tools for proper access of risk to human health is again a grand challenge.

1.4 TOXICITY OF NANOPARTICLES

As a result of their extremely small size and dimensions, these nanoparticles, which give unique advantages, also load them with potential hazards very similar to the particulate matters. These particles are responsible for various respiratory, cardiovascular, and gastrointestinal (GI) disorders. Many experiments conducted on rats have shown that the intra-tracheal installation of carbon nanotubes has resulted in an epithelioid granuloma, interstitial inflammation, peribronchial inflammation, and necrosis of lung. In the future, nanoparticle mediated delivery can provide an alternative route that can outwit the blood-brain barrier (BBB) but can also result in inflammatory responses in the brain which need to be evaluated.

The toxicity of nanoparticles can be frequently seen in GI systems, mainly in inflammatory bowel disease. If ingested, these nanoparticles can reach the blood and then can get circulated to various body parts and systems, resulting in toxicity.

1.5 CONCLUSION

Nanobiotechnology will certainly provide ways and means for developing new materials and methods that will enhance our ability to develop faster, more reliable, and more sensitive analytical systems. However, it is important to mention here that gradually, new discoveries will be seen in molecular recognition. The technique involved in the development of smaller and very smaller structures is due to advancements in fabrication techniques of semiconductor industries. This very small object will definitely find its applications beyond electronic devices, which were routine until now. The latest technologies in molecular detection applications are QDs and AuNPs. With this progression pace, future applications of this field would be *in vivo* sensors, and these are nanosize devices that could be injected, implanted, or ingested into our body. Such devices would have capabilities of sensing as well as transmitting information in the form of data to an externally linked device. These devices would be a great help to keep a constant vigil on the health of a person. To date, there is no such regulation for the application of such devices. However, it is possible to have some regulations for that.

ACKNOWLEDGMENTS

The authors are grateful to Dr. A. K. Chauhan, Founder President, and Dr. Aseem Chauhan, Chancellor Amity University Haryana, and Chairperson Amity Lucknow for providing us necessary facilities and support. We would also like to extend our gratitude to Maj. Gen. K. K. Ohri, AVSM (Retd.), Pro-Vice-Chancellor, Amity University, Uttar Pradesh Lucknow Campus for his constant support and encouragement.

KEYWORDS

- **gene therapy**
- **gold nanoparticles**
- **nanobiotechnology**
- **pathogen**
- **quantum dots**
- **tissue engineering**
- **toxicity**

REFERENCES

1. Abu-Youssef, M. A., Soliman, S. M., Langer, V., Gohar, Y. M., Hasanen, A. A., Makhyoun, M. A., Zaky, A. H., & Ohrstrom, L. R., (2010). Synthesis, crystal structure, quantum chemical calculations, DNA interactions, and antimicrobial activity of Ag (2-amino-3-methylpyridine) $2NO_3$ and Ag pyridine- 2-carboxaldoxime NO_3. *Inorg. Chem., 49*(21), 9788–9797.
2. Amna, T., Hassan, M. S., Nam, K. T., Bing, Y. Y., Barakat, N. A., Khil, M. S., & Kim, H. Y., (2012). Preparation, characterization, and cytotoxicity of CPT/Fe_2O_3-embedded PLGA ultrafine composite fibers: A synergistic approach to develop promising anticancer material. *Int. J. Nanomedicine, 7*, 1659–1670.
3. Ariga, T., (2006). Gene therapy for primary immunodeficiency diseases: Recent progress and misgivings. *Curr. Pharm. Des., 12*, 549–256.
4. Atala, A., (2005). Technology insight: Applications of tissue engineering and biological substitutes in urology. *Nat. Clin. Pract. Urol., 2*, 143.
5. Baker, B. M., Nathan, A. S., Gee, A. O., & Mauck, R. L., (2010). The influence of an aligned nanofibrous topography on human mesenchymal stem cell fibrochondrogenesis. *Biomaterials, 31*(24), 6190–6200.
6. Bhattacharyya, D., Singh, S., Satnalika, N., Khandelwa, A., & Jeon, S. H., (2009). Big things from a tiny world: A review. *International Journal of Science and Technology, 2*, 29–38.
7. Blanpain, C., (2010). Stem cells: Skin regeneration and repair. *Nature, 464*, 686–687.
8. Bruchez, M., Moronne, M., Gin, P., Weiss, S., & Alivisatos, A. P., (1998). Semiconductor nanocrystals as fluorescent biological labels. *Science, 281*, 2013–2016.
9. Cavicchioli, M., Massabni, A. C., Heinrich, T. A., Costa-Neto, C. M., Abrão, E. P., & Fonseca, B. A. L., (2010). Pt[II] and Ag[I] complexes with acesulfame: Crystal structure and a study of their antitumoral, antimicrobial and antiviral activities. *J. Inorg. Biochem., 104*, 533–540.
10. Chahardeh, J. B., (2012). A review of graphene transistors. *International Journal of Advance Research in Computer and Communication Engineering, 1*, 193–197.
11. Chakraborty, S., Liao, I. C., Adler, A., & Leong, K. W., (2009). Electrohydrodynamics: A facile technique to fabricate drug delivery systems. *Adv. Drug Deliv. Rev., 61*, 1043–1054.
12. Chan, W. C. W., & Nie, S. M., (1998). Quantum dot bioconjugates for ultrasensitive nonisotopic detection. *Science, 281*, 2016–2018.
13. Chandrasekaran, A. R., Venugopal, J., Sundarrajan, S., & Ramakrishna, S., (2011). Fabrication of a nanofibrous scaffold with improved bioactivity for culture of human dental fibroblasts for skin regeneration. *Biomed. Mater., 6*(1), 15. doi: 10.1088/1748–6041/6/1/015001.
14. Cream, C., Lahiff, E., Gilmartin, N., Diamond, D., & O'Kennedy R., (2011). Polyaniline nanofibers as templates for the covalent immobilization of biomolecules. *Synthetic Metals, 161*(3&4), 285–292.
15. Cunha, C., Panseri, S., & Antonini, S., (2011). Emerging nanotechnology approaches in tissue engineering for peripheral nerve regeneration. *Nanomedicine, 7*, 50–59.

16. De La Isla, A., Brostow, W., Bujard, B., Estevez, M., Rodriguez, J. R., Vargas, S., & Castano, V. M., (2003). Nanohybrid scratch resistant coating for teeth and bone viscoelasticity manifested in tribology. *Mat. Resr. Innovat., 7*, 110–114.
17. Edelstein, R. L., Tamanaha, C. R., Sheehan, P. E., Miller, M. M., Baselt, D. R., Whitman, L. J., & Colton, R. J., (2000). The BARC biosensor applied to the detection of biological warfare agents. *Biosensors Bioelectron, 14*, 805–813.
18. Ehrenberg, R., (2013). DNA could soon prove practical for long-term data storage. *Science News, 1*, 5–6.
19. Elbaz, J., Lioubashevski, O., Wang, F., Remacle, F., Levine, R. D., & Willner, I., (2010). DNA computing circuits using libraries of DNAzymen subunits. *Nature Nanotechnology, 5*, 417–422.
20. Fakruddin, M., Hossain, Z., & Afroz, H., (2012). Prospects and applications of nanobiotechnology: A medical perspective. *J. Nanobiotechnol., 10*, 31.
21. Fine, E. G., Valentini, R. F., & Aebischer, P., (2000). In: Lanza, R. P., Langer, R., & Vacanti, J., (eds.), Nerve regeneration, principles of tissue engineering. Academic Press, San Diego.
22. Fouad, M., Kaji, N., Jabasini, M., Tokeshi, M., & Baba, Y., (2008). *Nanotechnology Meets Plant Biotechnology: Carbon Nanotubes Deliver DNA and Incorporate Into the Plant Cell Structure*. Twelfth International Conference on Miniaturized Systems for Chemistry and Life Sciences, San Diego, California, USA.
23. Frankel, F. C., & Whitesides, G. M., (2009). *No Small Matter: Science on the Nanoscale*. Cambridge, Massachusetts: The Belknap Press of Harvard University Press.
24. Giannoudis, P. V., Dinopoulos, H., & Tsiridis, E., (2005). Bone substitution: An update. *Injury, 36*(3), 520–527.
25. Gilmartin, N., & O'Kennedy, R., (2012). Nanobiotechnologies for the detection and reduction of pathogens. *Enzyme Microb. Technol., 50*, 87–95.
26. Greiner, A., & Wendorff, J. H., (2007). Electrospinning: A fascinating method for the preparation of ultrathin fibers. *Chem. Int. Ed. Engl., 46*(30), 5670–5703.
27. Guo, P., Shu, Y., Binzel, D., & Cinier, M., (2012). Synthesis, conjugation, and labeling of multifunctional pRNA nanoparticles for specific delivery of siRNA, drugs, and other therapeutics to target cells. *Methods Mol. Biol., 928*, 197–219.
28. Hanakawa, Y., Shirakata, Y., Nagai, H., Yahata, Y., Tokumaru, S., Yamasaki, K., Tohyama, M., Sayam, K., & Hashimoto, K., (2005). Cre-loxP adenovirus-mediated foreign gene expression in skin-equivalent keratinocytes. *Br. J. Dermatol., 152*, 1391–1392.
29. Jain, K. K., (2005). The role of nanobiotechnology in drug discovery. *Drug Discov. Today, 10*, 1435–1442.
30. Jin, G., Prabhakaran, M. P., Kai, D., Annamalai, S. K., Arunachalam, K. D., & Ramakrishna, S., (2013). Tissue engineered plant extracts as nanofibrous wound dressing. *Biomaterials, 34*(3), 724–734.
31. Jyoti, A., Pandey, P., Singh, S. P., Jain, S. K., & Shanker, R., (2010). Colorimetric detection of nucleic acid signature of shiga toxin producing Escherichia coli using gold nanoparticles. *J. Nanosci. Nanotechnol., 10*, 4154–4158.
32. Kostoff, R. N., Koytcheff, R. G., & Lau, C. G. Y., (2007). Global nanotechnology research literature overview. *Technological Forecasting and Social Change, 74*, 1733–1747.

33. Kompella, U. B., Amrite, A. C., Ravi, R. P., & Durazo, S. A., (2013). Nanomedicines for back of the eye drug delivery, gene delivery and imaging. *Prog. Retin. Eye Res., 36*, 172–198.
34. Lahiff, E., Lynam, C., Gilmartin, N., Kennedy, R., & O'Diamond, D., (2010). The increasing importance of carbon nanotube and nanostructured conducting polymers in biosensors. *Anal. Bioanal. Chem., 398*(4), 1575–1589.
35. Langer, R., (1999). Selected advances in drug delivery and tissue engineering. *J. Control Release, 62*(91&92), 7–11.
36. Lee, S. H., & Shin, H., (2007). Matrices and scaffolds for delivery of bioactive molecules in bone and cartilage tissue engineering. *Adv. Drug Deliv. Rev., 59*(4&5), 339–359.
37. Ma, J., Wong, H., Kong, L. B., & Peng, K. W., (2003). Biomimetic processing of nanocrystallite bioactive apatite coating on titanium. *Nanotechnology, 14*, 619–623.
38. Mah, C., Zolotukhin, I., Fraites, T. J., Dobson, J., Batich, C., & Byrne, B. J., (2000). Microsphere-mediated delivery of recombinant AAV vectors in vitro and *in vivo*. *Mol Therapy, 1*, S239.
39. Mahtab, R., Rogers, J. P., & Murphy, C. J., (1995). Protein-sized quantum dot luminescence can distinguish between "straight," "bent," and "kinked" oligonucleotides. *J. Am. Chem. Soc., 117*, 9099–9100.
40. Martin, P. (1997). Wound healing-aiming for perfect skin regeneration. *Science, 276*(5309), 75–81.
41. Mashaghi, S., Jadidi, T., Koenderink, G., & Mashaghi, A., (2013). Lipid nanotechnology. *Int. J. Mol. Sci., 14*, 4242–4282.
42. Matteo, C., Candido, P., Vera, R., & Fransisca, V., (2012). Current and future nanotech applications in oil industry. *American Journal of Applied Sciences, 9*, 784–793.
43. Middelkoop, E., Bogaerdt, A. J. V. D., Lamme, E. N., Hoekstra, M. J., Brandsma, K., & Ulrich, M. M., (2004). Procine wound models for skin substitution and burn treatment. *Biomaterials, 25*(9), 1559–1567.
44. Mohamed, A., & Xing, M. M., (2012). Nanomaterials and nanotechnology for skin tissue engineering. *Int. J. Burns Trauma, 2*(1) 29–41.
45. Molday, R. S., & MacKenzie, D., (1982). Immunospecific ferromagnetic iron dextran reagents for the labeling and magnetic separation of cells. *J. Immunol. Methods, 52*, 353–367.
46. Morais, M. G., Martins, V. G., Steffens, D., Pranke, P., & Viera Da, C. J. A., (2014). Biological applications of nanobiotechnology. *J. Nanosciences and Nanotechnology, 14*, 1007–1017.
47. Nam, J. M., Thaxton, C. C., & Mirkin, C. A., (2003). Nanoparticles-based bio-bar codes for the ultrasensitive detection of proteins. *Science, 301*, 1884–1886.
48. Niedzwiecki, D. J., Iyer, R., Borer, P. N., & Movilenau, L., (2013). Sampling a biomarker of the human immunodeficiency virus across a synthetic nanopore. *ACS Nano., 7*, 3341–3350.
49. Oliver J, D., (2010). Recent findings on the viable but nonculturable state in pathogenic bacteria. *FEMS Microbiol. Rev., 34*, 415–425.
50. Panatarotto, D., Prtidos, C. D., Hoebeke, J., Brown, F., Kramer, E., Briand, J. P., Muller, S., Prato, M., & Bianco, A., (2003). Immunization with peptide-functionalized

carbon nanotubes enhances virus-specific neutralizing antibody responses. *Chemistry and Biology, 10*, 961–966.
51. Paolo, F., Kricka, L. J., Surrey, S., & Rodzinski, P., (2005). Nanobiotechnology: The promise and reality of new approaches to molecular recognition. *Trends in Biotechnology, 23*(4), 2005.
52. Parak, W. J., Boudreau, R., Gros, M. L., Gerion, D., Zanchet, D., Micheel, C. M., Williams, S. C., Alivisatos, A. P., & Larabell, C. A., (2002). Cell motility and metastatic potential studies based on quantum dot imaging of phagokinetic tracks. *Adv. Mater., 14*, 882–885.
53. Qi, H., Hu, P., Xu, J., & Wang, A., (2006). Encapsulation of drug reservoirs in fibers by emulsion electrospinning: Morphology characterization and preliminary release assessment. *Biomacromolecules, 7*, 2327–2330.
54. Sahoo, S. K., Parveen, S., & Panda J, J., (2007). The present and future of nanotechnology in human health care. *Nanomedicine, 3*, 20–31.
55. Saiz, E., Zimmermann, E. A., Lee, J. S., Wegst, U. G., & Tomsia, A. P., (2013). Perspectives on the role of nanotechnology in bone tissue engineering. *Dent. Mater., 29*, 103–115.
56. Salata, O. V., (2004). Applications of nanoparticles in biology and medicine. *Journal of Nanobiotechnology, 2*, 3.
57. Sasmazel, H. T., (2011). Novel hybrid scaffolds for the cultivation of osteoblast cells. *Int. J. Biol. Macromol., 49*(4), 838–846.
58. Shi, J., Votruba, A. R., Farokhzad, O. C., & Langer, R., (2010). Nanotechnology in drug delivery and tissue engineering: From discovery to applications. *Nano Lett., 10*, 3223–3230.
59. Shu, Y., Cinier, M., Shu, D., & Guo, P., (2011). Assembly of multifunctional phi29 pRNA nanoparticles for specific delivery of siRNA and other therapeutics to targeted cells. *Methods, 54*, 204–214.
60. Shu, Y., Haque, F., Shu, D., Li, W., Zhu, Z., Kotb, M., Lyubchenko, Y., & Guo, P., (2013). Fabrication of 14 different RNA nanoparticles for specific tumor targeting without accumulation in normal organs. *RNA, 19*, 767–777.
61. Singh, R., & Nalwa, H. S., (2011). Medical applications of nanoparticles in biological imaging, cell labeling, antimicrobial agents, and anticancer nanodrugs. *Biomed. Nanotechnol., 7*, 489–503.
62. Song, W., Markel, D. C., Wang, S., Shi, T., Mao, G., & Ren, W., (2012). Electrospun polyvinyl alcohol collagen hydroxyapatite nanofibers: A biomimetic extracellular matrix for osteoblastic cells. *Nanotechnology, 23*(11), 115101.
63. Steffens, D., Lersch, M., Rosa, A., Scher, C., Crestani, T., Morais, M. G., Costa, J. A., & Pranke, P., (2013). A new biomaterial of nanofibers with the microalga Spirulina as scaffolds to cultivate with stem cells for use in tissue engineering. *J. Biomed. Nanotechnol., 9*(4), 710–718.
64. Subramanian, A., Krishnan, U. M., & Sethuraman, S., (2011). Fabrication of uniaxially aligned 3D electrospun scaffolds for neural regeneration. *Biomed. Mater., 6*(2), 025004.
65. Suh, W. H., Suslick, K. S., Stucky, G. D., & Suh, Y. H., (2009). Nanotechnology, nanotoxicology, and neuroscience. *Prog. Neurobiol. 87*, 133–170.

66. Vasita, R., & Katti, D. S., (2006). Nanofibers and their applications in tissue engineering. *Int. J. Nanomedicine, 1*, 15–30.
67. Venugopal, J. R., Prabhakaran, M. P., Mukherjee, S., Ravichandran, R., Dan, K., & Ramakrishna, S., (2012). Biomaterial strategies for alleviation of myocardial infarction. *J. R. Soc. Interface, 9*(66), 1–19.
68. Wang, S., Mamedova, N., Kotov, N. A., Chen, W., & Studer, J., (2002). Antigen/antibody immunocomplex from CdTe nanoparticle bioconjugates. *Nano Letters, 2*, 817–822.
69. Wang, Y., Ravindranath, S., & Irudayaraj, J., (2011). Seperation and detection of multiple pathogens in a food matrix by magnetic SERS nanoprobes. *Anal. Bioanal. Chem., 399*(3), 1271–1278.
70. Weissleder, R., Elizondo, G., Wittenburg, J., Rabito, C. A., Bengele, H. H., & Josephson, L., (1990). Ultra small superparamagnetic iron oxide: Characterization of a new class of contrast agents for MR imaging. *Radiology, 175*, 489–493.
71. Wepener, I., Richter, W., Papendorp, V. D., & Joubert, A. M., (2012). In vitro osteoclast like and osteoblast cells response to electrospun calcium phosphate biphasic candidate scaffolds for bone tissue engineering. *J. Mater. Sci. Mater., Med., 23*(12), 3029–3040.
72. Wheeldon, I., Farhadi, A., Bick, A. G., Jabbari, E., & Khademhosseini, A., (2011). Nanoscale tissue engineering: Spatial control over cell- materials interactions. *Nanotechnology, 22*(21), 212, doi: 10.1088/0957.
73. Whitesides, G. M., (2003). The 'right' size in nanobiotechnology. *Nature Biotechnology, 21*, 1161–1165.
74. Yamaguchi, R., Takami, Y., Yamaguchi, Y., & Shimazaki, S., (2007). Bone marrow-derived myofibroblasts recrited to the upper dermis appear beneath regenerating epidermis after deep dermal burn injury. *Wound Repair Regen., 15*(1), 87–93.
75. Yoshida, J., & Kobayashi, T., (1999). Intracellular hyperthermia for cancer using magnetite cationic liposomes. *J. Magn. Magn. Mater., 194*, 176–184.
76. Zhou, J., Shu, Y., Guo, P., Smith, D. D., & Rossi, J. J., (2011). Dual functional RNA nanoparticles containing phi29 motor pRNA and anti-gp120 aptamer for cell-type specific delivery and HIV-1 inhibition. *Methods, 54*, 284–294.
77. Zuijlen, P. P., Van Trier, A. J., Van Vloemans, J. F., Groenevelt, F., Kreis, R. W., & Middelkoop, E., (2000). Graft survival and effectiveness of dermal substitution in burns and reconstructive surgery in a one-stage grafting model. *Plast. Reconstr. Surg., 106*(3), 615–623.
78. Aaron, J., Nitin, N., Travis, K., Kumar, S., Collier, T., Park, S., Y., Jose, M., Coghlan, L., Follen, M., Richards, K. R., & Sokolov, K., (2007). Plasmon resonance coupling of metal nanoparticles for molecular imaging of carcinogenesis in vivo. *J Biomed Opt. 12*(3), 034007.

CHAPTER 2

Nanomaterials: Novel Preparation Routes, Characterizations, and Applications

JUHI SAXENA[1] and ANUPAM JYOTI[2*]

[1]*Dr. B. Lal Institute of Biotechnology, Malviya Industrial Area, Jaipur, India*

[2]*Amity Institute of Biotechnology, Amity University Rajasthan, Jaipur, India*

[*]*Corresponding author. E-mail: ajyoti@jpr.amity.edu; anupamjyoti@rediffmail.com*

ABSTRACT

Nanotechnology is a multidisciplinary area as it finds diverse applications in biomedicine, catalysis, molecular detection, and many more. Scientists are facing challenges in obtaining nanoparticles (NPs) with high mono-dispersity, specific composition, and size. In the context of this, several research groups have exploited the synthesis of NPs by the biological system over non-biological systems due to many reasons. This chapter intends to present the biosynthesis of NPs involving the use of bacteria, yeast, fungi, and plants. Furthermore, we discussed the different characterization methods and overviewed on potential applications of nanomaterials in different fields.

2.1 INTRODUCTION

Nanotechnology is a modern era of technology that influences all aspects of human life. Scientists are interested in the synthesis of nanoparticles (NPs) with high mono-dispersity, specific composition, and size, which form the core part of the nanomaterials [1]. Several methods, including physical, chemical, and biological approaches, have been employed for the synthesis of NPs. The physical and chemical approaches are complicated, outdated, costly, and produce hazardous toxic wastes that are harmful to the environment and human health. Biogenic synthesis utilizing enzymes secreted by bacteria, fungi, yeast, and plants for NPs synthesis are gaining more importance [2]. NPs have important applications in therapeutics as it fights against pathogenic microbes and hence have essential to clinical application. NPs kill the microbes by penetrating the cell wall, damaging the cell membrane, interacting with respiratory enzymes, inhibiting various signaling pathways, producing excessive free radicals, and damaging the DNA [3].

2.1.1 BIOLOGICAL SYNTHESIS OF NANOPARTICLES (NPS)

The biological synthesis of NPs is preferred over physical and chemical means because of its cost-effectiveness, rapid synthesis, the option of size and shape control, less toxic, and eco-friendly approach. In general, extracts of different biological sources like bacteria, fungi, yeast, and plants containing natural reducing agents are added to metal ion solutions and observed for a color change that results in NPs synthesis (Figure 2.1). In addition, these biomolecules also confer stability to NPs.

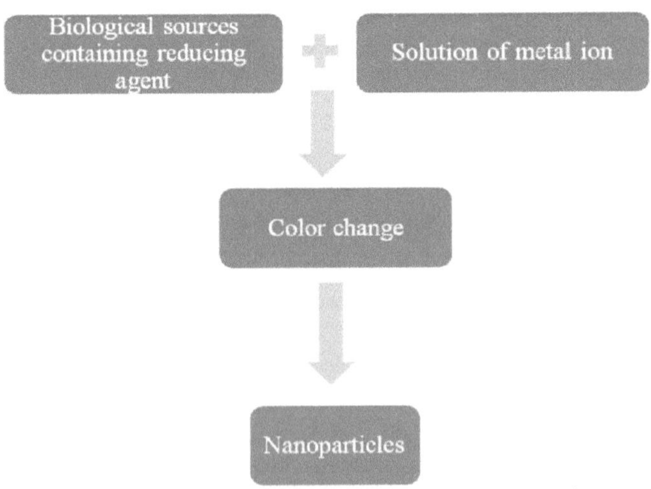

FIGURE 2.1 Schematics of nanoparticle synthesis from biological sources.

2.1.1.1 NANOPARTICLE (NP) SYNTHESIS BY BACTERIA

In previous years, the synthesis of NPs using bacteria has gained immense importance. Bacteria secrete various enzymes, including nitrate reductase that reduces metallic salts into NPs. In doing this, the size of metal reduced, and hence, surface area to volume ratio increases. Several bacterial species, including *Escherichia coli* [4], *Ureibacillus thermosphaericus* [5], *Staphylococcus aureus* [6], *Corynebacterium* strain SH09 [7], *Bacillus cereus* [8] and many more have been shown to synthesize NPs. The understanding of natural processes will apparently help in the discovery of an entirely new and unexplored methodology of metal NP synthesis.

2.1.1.2 NANOPARTICLE (NP) SYNTHESIS BY PLANT EXTRACTS

NPs synthesis by plants is gaining importance as it provides a single step biosynthesis process. Plants offer a simple source of NP synthesis, which is free from any toxicants. Stability of NPs is an important issue, which is resolved by plant-mediated synthesis as it provides natural capping agent. Furthermore, plant-mediated NP synthesis is a cost-effective and environment-friendly approach. Plants secrete reductase enzyme extracellularly that helps in the reduction of metal ions into NPs hence synthesized extracellularly. Primarily gold and silver nanoparticles (AgNPs) have been shown to synthesize by plant extracts. NP synthesis furthermore carried out using *Azadirachta indica* [9], *Mirabilis jalapa* [10], *Murraya koenigii* [11], *Cardiospermum helicacabum* [12], *Aloe vera* [13], and many more.

2.1.1.3 NANOPARTICLE (NP) SYNTHESIS BY FUNGI

Biological production of NPs by fungi is catching attention from the researchers nowadays. Different species of fungi like [14, 15] *Fusarium solani* [16], *Pleurotus sajorcaju* [17], *Fusarium semitectum* [18], *Alternaria alternata* [19], *Fusarium acuminatum* [20], *Penicillium fellutanum* [21], *Penicillium brevicompactum* [22], *Aspergillus clavatus* [23], and *Sclerotinia sclerotiorum* [24] have been known for synthesis of NPs. Fungi have an upper edge in NP synthesis over other sources because of higher bioaccumulation, comparatively economic, effortless synthesis method, an excellent source of various extracellular enzymes, and simple downstream processing and biomass handling. There are various reasons onto which fungi can be preferred over bacteria and plants in NP synthesis [25]:

1. **Large Secretor of Protein:** Fungi have the capacity to produce large amounts of extracellular enzymes that helps in the reduction of metal ions into NPs.
2. **Ease in Isolation**: Being simple nutritional requirements fungi are easy to isolate and subculture that further helps in NPs synthesis.
3. **Extracellular Synthesis:** Being an excellent secretor of enzymes outside the cell, fungi synthesize NP extracellularly that is helpful in easier downstream processing.

2.1.1.4 NANOPARTICLE (NP) SYNTHESIS BY YEAST

Yeast has also been shown to synthesize NPs with simple downstream processing. Biosynthesis of NPs using *Candida glabrata* and *Schizosaccharomyces pombe* [26], *Rhodosporidium diobovatum* [27], *Saccharomyces cerevisiae,* and *Cryptococcus humicola* has been documented [28].

2.2 CHARACTERIZATION

There are various attributes for NP characterization. These include size, morphology, and surface charge. A range of diverse techniques like spectroscopy techniques (UV-vis spectroscopy, Fourier transform infrared (FTIR) spectroscopy), microscopy techniques (scanning electron microscopy, transmission electron microscopy (TEM)), x-ray based characterization techniques [x-ray photoemission spectroscopy (XPS), energy dispersive x-ray analysis (EDX), and particles size analyzer (dynamic light scattering (DLS)] are handful to characterize NPs (Figure 2.2).

1. **UV-vis spectroscopy:** This technique characterizes NPs based on the color they produce. There is free movement of electrons in metallic NPs. Hence, when the oscillation of electrons of NPs is in resonance with the light wave, oscillations result in a unique color and produces surface plasmon resonance (SPR) absorption band [29].

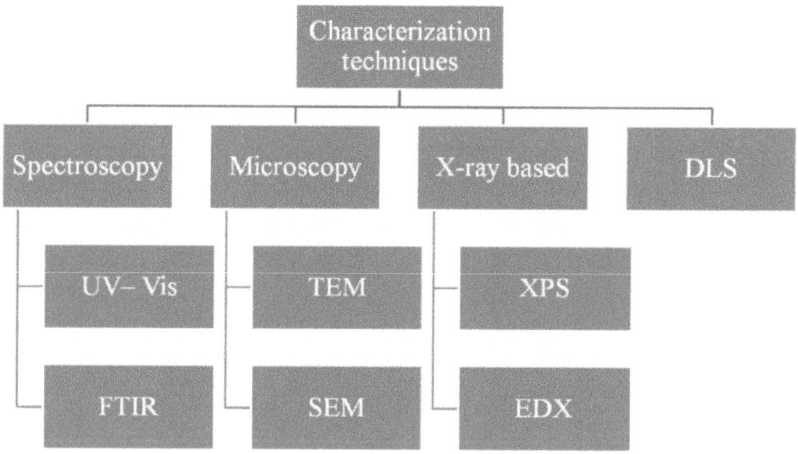

FIGURE 2.2 Characterization techniques of nanoparticles.

2. **Fourier Transform Infrared (FTIR):** In this technique, NPs are analyzed by observing the atomic vibrations when exposed to electromagnetic radiation in the wavelength ranging from 4000–400 cm^{-1}. This provides the information which can assist molecular interaction studies such as hydrogen bonding, conformational changes, amino acid functionalization and also confirm the secondary structure of the proteins on the basis of the absorption of amide bonds [30].
3. **Transmission Electron Microscopy (TEM):** It is a powerful and technique for determining the structure and absolute size of NPs. Samples prepared for the TEM analysis must be very thin and able to bear up the high vacuum present inside the instrument. These samples are fixed with uranyl acetate, and a beam of electrons is transmitted through ultra-thin sample results in surface characteristics [31].
4. **Scanning Electron Microscopy (SEM):** It provides three-dimensional images of the NPs. For SEM characterization, dried powder form of NP is taken and then mounted on a sample holder followed by coating with gold. Furthermore, a beam of electron scans the sample, and the surface characteristics of the sample are obtained from the secondary electrons emitted from the sample surface [31].
5. **X-Ray Photoemission Spectroscopy (XPS):** It is a quantitative spectroscopic technique that is used to estimate the elemental composition [32].

6. **Energy Dispersive X-Ray Analysis (EDX):** It is a useful technique performed along with SEM. In this technique, an electron beam strikes with the surface of the specimen results in X-ray to be emitted from the material. The energy of these emitted X-rays depends on the type of elements present in the material.
7. **Dynamic Light Scattering (DLS):** It is the fastest method of determining particle size. DLS is used to determine the size of NPs in colloidal suspensions. When a monochromatic light hits a solution of NPs in Brownian motion, it causes a change in the wavelength of incoming light due to the Doppler shift. This change is related to the size of the particle [33].

2.3 NANOMATERIALS TOXICITY

Nanomaterials with their unique properties have been exploited the most in various industrial sectors; however, evaluation of their biological, as well as environmental toxicity, is a major area of concern. *In vivo* and *in vitro* models, including plant, microbial systems, animals, primary cells, and cell lines have been tested for increasing accumulation nanomaterials in environmental systems [34]. Nanomaterial for their therapeutic applications should have no or less cytotoxicity. Following administration, any exogenous agent interacts with blood cells, which are independent from the route of administration, therefore the interaction with blood components needed to be studied for cytotoxicity studies [35]. Hemolysis assay can be employed for estimating toxicity on red blood cells. The assay revealed safe rates of RBC lysis of under 5% (1.1 to 4.6%) [36]. Mechanistically, NPs produce reactive oxygen species and led to genotoxicity [37]. NPs have also been reported to induce an inflammatory response through the activation of various cytokines [38].

2.4 APPLICATIONS

NPs find diverse applications in various fields such as cosmetics, agriculture, pharmaceuticals, food and beverages, polymers, etc. [39] (Figure 2.3).

1. **Anti-Bacterial:** Emergence of multi-drug resistant bacterial strains is increasing day by day and representing a major threat. NPs, especially silver NP, are remarkable in killing both Gram-negative and Gram-positive bacterial pathogens. Mechanistically,

NPs enter to the bacterial cells by perforation in the cell wall, degrading membrane proteins, inhibiting signaling molecules, damaging DNA, and producing excessive free radicals.

2. **Biosensors:** Being specific electrochemical properties, NPs are instrumental in biosensor development. These biosensors are highly specific and sensitive and generate a response in a quick time. NPs provide a platform for biomolecules to adhere. The physicochemical properties of NPs are suitable for designing new sensing devices.
3. **Disease Diagnosis:** NPs have also found applications in the diagnosis of biomedical disorders. Presently, imaging techniques using iron oxide nanoparticles (IONPs) are used for the diagnosis of diseases.

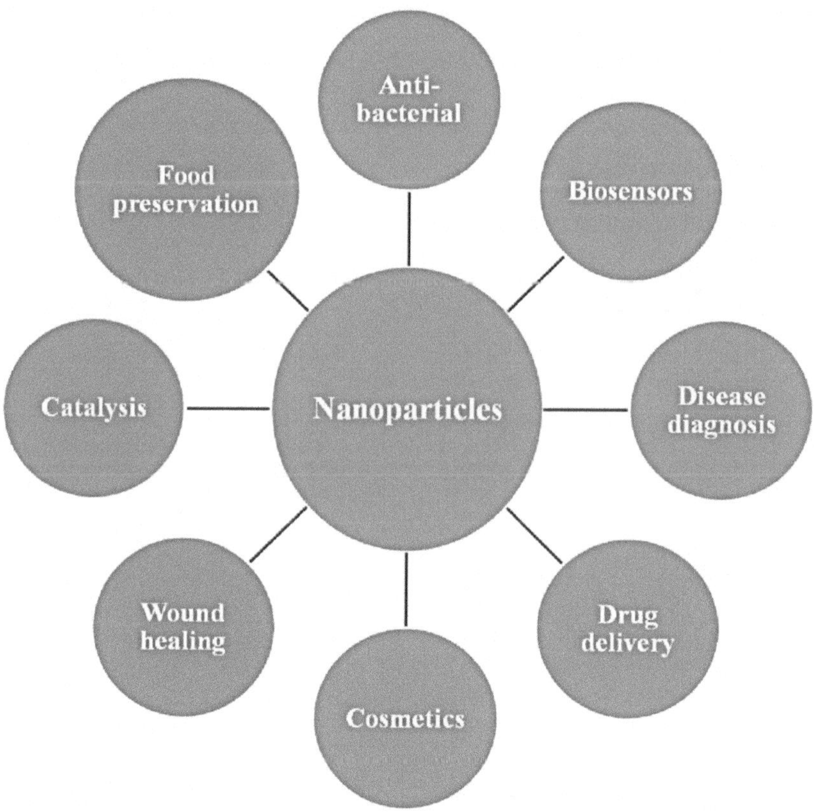

FIGURE 2.3 Applications of nanotechnology.

4. **Drug Delivery System:** One of the important applications of NPs is in delivering the drug. NPs carry the drug and deliver it at the site of alignment. NPs improve important features, including solubility, stability, kinetic property, availability, and enhancing their efficacy.
5. **Cosmetics:** NPs are widely used in cosmetics. Zinc oxide and titanium dioxide NPs are used in various products, including sunscreens, cosmetics, sanitizer, and toothpaste.
6. **Wound Healing:** NPs are used in healing wounds as compared to standard silver sulfadiazine ointment.
7. **Catalysis:** NPs have remarkable biocatalytic properties. Metal NPs are used in the immobilization of enzymes for enhancing enzymatic activity.
8. **Vegetable and Food Preservation:** NPs have been shown to improve the shelf life of vegetables and fruits when added into sodium alginate thin film.

KEYWORDS

- **Fourier transform infrared**
- **nanoparticles**
- **scanning electron microscopy**
- **surface plasmon resonance**
- **transmission electron microscopy**
- **x-ray photoemission spectroscopy**

REFERENCES

1. Kashyap, P. L., Kumar, S., Srivastava, A. K., & Sharma, A. K., (2013). Myconanotechnology in agriculture: A perspective. *World Journal of Microbiology and Biotechnology*, *29*(2), 191–207.
2. Kowshik, M., Deshmukh, N., Vogel, W., Urban, J., Kulkarni, S. K., & Paknikar, K. M., (2002). Microbial synthesis of semiconductor CdS nanoparticles, their characterization, and their use in the fabrication of an ideal diode. *Biotechnology and Bioengineering*, *78*(5), 583–588.

3. Azam, A., Ahmed, A. S., Oves, M., Khan, M. S., Habib, S. S., & Memic, A., (2012). Antimicrobial activity of metal oxide nanoparticles against gram-positive and gram-negative bacteria: A comparative study. *International Journal of Nanomedicine, 7*, 6003.
4. Gurunathan, S., Kalishwaralal, K., Vaidyanathan, R., Venkataraman, D., Pandian, S. R. K., Muniyandi, J., & Eom, S. H., (2009). Biosynthesis, purification and characterization of silver nanoparticles using *Escherichia coli*. *Colloids and Surfaces B: Biointerfaces, 74*(1), 328–335.
5. Juibari, M. M., Abbasalizadeh, S., Jouzani, G. S., & Noruzi, M., (2011). Intensified biosynthesis of silver nanoparticles using a native extremophilic *Ureibacillus thermosphaericus* strain. *Materials Letters, 65*(6), 1014–1017.
6. Nanda, A., & Saravanan, M., (2009). Biosynthesis of silver nanoparticles from *Staphylococcus aureus* and its antimicrobial activity against MRSA and MRSE. *Nanomedicine: Nanotechnology, Biology and Medicine, 5*(4), 452–456.
7. Zhang, H., Li, Q., Lu, Y., Sun, D., Lin, X., Deng, X., & Zheng, S., (2005). Biosorption and bioreduction of diamine silver complex by Corynebacterium. *Journal of Chemical Technology and Biotechnology, 80*(3), 285–290.
8. Babu, M. G., & Gunasekaran, P., (2009). Production and structural characterization of crystalline silver nanoparticles from Bacillus cereus isolate. *Colloids and Surfaces B: Biointerfaces, 74*(1), 191–195.
9. Shankar, S. S., Rai, A., Ahmad, A., & Sastry, M., (2004). Rapid synthesis of Au, Ag, and bimetallic Au core—Ag shell nanoparticles using Neem (*Azadirachta indica*) leaf broth. *Journal of Colloid and Interface Science, 275*(2), 496–502.
10. Vankar, P. S., & Bajpai, D. (2010). Preparation of gold nanoparticles from Mirabilis jalapa flowers. *Indian Journal of Biochemistry and Biophysics, 47*, 157–160.
11. Christensen, L., Vivekanandhan, S., Misra, M., & Mohanty, A. K., (2011). Biosynthesis of silver nanoparticles using *Murraya koenigii* (curry leaf): An investigation on the effect of broth concentration in reduction mechanism and particle size. *Adv. Mat. Lett., 2*(6), 429–434.
12. Mitra, B., Vishnudas, D., Sant, S. B., & Annamalai, A., (2012). Green-synthesis and characterization of silver nanoparticles by aqueous leaf extracts of *Cardiospermum helicacabum* leaves. *Drug Invention Today, 4*(2), 342–344.
13. Chandran, S. P., Chaudhary, M., Pasricha, R., Ahmad, A., & Sastry, M., (2006). Synthesis of gold nanotriangles and silver nanoparticles using Aloe vera plant extract. *Biotechnology Progress, 22*(2), 577–583.
14. Karbasian, M., Atyabi, S. M., Siadat, S. D., Momen, S. B., & Norouzian, D., (2008). Optimizing nano-silver formation by Fusarium oxysporum PTCC 5115 employing response surface methodology. *American Journal of Agricultural and Biological Science, 3*(1), 433–437.
15. Ahmad, A., Mukherjee, P., Senapati, S., Mandal, D., Khan, M. I., Kumar, R., & Sastry, M., (2003). Extracellular biosynthesis of silver nanoparticles using the fungus *Fusarium oxysporum*. *Colloids and Surfaces B: Biointerfaces, 28*(4), 313–318.
16. Ingle, A., Rai, M., Gade, A., & Bawaskar, M., (2009). *Fusarium solani*: A novel biological agent for the extracellular synthesis of silver nanoparticles. *Journal of Nanoparticle Research, 11*(8), 2079–2085.

17. Nithya, R., & Ragunathan, R., (2009). Synthesis of silver nanoparticle using *Pleurotus sajor caju* and its antimicrobial study. *Digest Journal of Nanomaterials and Biostructures*, *4*(4), 623–629.
18. Bhainsa, K. C., & D'souza, S. F., (2006). Extracellular biosynthesis of silver nanoparticles using the fungus *Aspergillus fumigatus*. *Colloids and Surfaces B: Biointerfaces*, *47*(2), 160–164.
19. Gajbhiye, M., Kesharwani, J., Ingle, A., Gade, A., & Rai, M., (2009). Fungus-mediated synthesis of silver nanoparticles and their activity against pathogenic fungi in combination with fluconazole. *Nanomedicine: Nanotechnology, Biology and Medicine*, *5*(4), 382–386.
20. Ingle, A., Gade, A., Pierrat, S., Sonnichsen, C., & Rai, M., (2008). Mycosynthesis of silver nanoparticles using the fungus *Fusarium acuminatum* and its activity against some human pathogenic bacteria. *Current Nanoscience*, *4*(2), 141–144.
21. Kathiresan, K., Manivannan, S., Nabeel, M. A., & Dhivya, B., (2009). Studies on silver nanoparticles synthesized by a marine fungus, *Penicillium fellutanum* isolated from coastal mangrove sediment. *Colloids and Surfaces B: Biointerfaces*, *71*(1), 133–137.
22. Shaligram, N. S., Bule, M., Bhambure, R., Singhal, R. S., Singh, S. K., Szakacs, G., & Pandey, A., (2009). Biosynthesis of silver nanoparticles using aqueous extract from the compactin producing fungal strain. *Process Biochemistry*, *44*(8), 939–943.
23. Verma, V. C., Kharwar, R. N., & Gange, A. C., (2010). Biosynthesis of antimicrobial silver nanoparticles by the endophytic fungus *Aspergillus clavatus*. *Nanomedicine*, *5*(1), 33–40.
24. Saxena, J., Sharma, P. K., Sharma, M. M., & Singh, A., (2016). Process optimization for green synthesis of silver nanoparticles by *Sclerotinia sclerotiorum*. *Springer Plus*, *5*(1), 1–10.
25. Rai, M., Yadav, A., Bridge, P., Gade, A., Rai, M., & Bridge, P. D., (2009). Myconanotechnology: A new and emerging science. *Applied Mycology*, 258–267.
26. Dameron, C. T., Reese, R. N., Mehra, R. K., Kortan, A. R., Carroll, P. J., Steigerwald, M. L., & Winge, D. R., (1989). Biosynthesis of Cadmium Sulphide Quantum Semiconductor Crystallites. *Nature, 338*, 596–597.
27. Seshadri, S., Saranya, K., & Kowshik, M., (2011). Green synthesis of lead sulfide nanoparticles by the lead resistant marine yeast, *Rhodosporidium diobovatum*. *Biotechnology Progress*, *27*(5), 1464–1469.
28. Vainshtein, M., Belova, N., Kulakovskaya, T., Suzina, N., & Sorokin, V., (2014). Synthesis of magneto-sensitive iron-containing nanoparticles by yeasts. *Journal of Industrial Microbiology and Biotechnology*, *41*(4), 657–663.
29. Mulvaney, P., (1996). Surface plasmon spectroscopy of nanosized metal particles. *Langmuir*, *12*(3), 788–800.
30. Barnett, N. W., Dent, G., & Chalmers, J. M., (1997). *Industrial Analysis with Vibrational Spectroscopy*.
31. Molpeceres, J., Aberturas, M. R., & Guzman, M., (2000). Biodegradable nanoparticles as a delivery system for cyclosporine: Preparation and characterization. *Journal of Microencapsulation*, *17*(5), 599–614.

32. Watts, J. F., & Wolstenholme, J., (2003). An introduction to surface analysis by XPS and AES. *An Introduction to Surface Analysis by XPS and AES.* by John, F. Watts, John Wolstenholme, pp. 224. ISBN 0–470–84713–1. Wiley-VCH, 224.
33. De Assis, D. N., Mosqueira, V. C. F., Vilela, J. M. C., Andrade, M. S., & Cardoso, V. N., (2008). Release profiles and morphological characterization by atomic force microscopy and photon correlation spectroscopy of 99m Technetium-fluconazole nanocapsules. *International Journal of Pharmaceutics*, *349*(1), 152–160.
34. Magdolenova, Z., Collins, A., Kumar, A., Dhawan, A., Stone, V., & Dusinska, M., (2014). Mechanisms of genotoxicity: A review of *in vitro* and in vivo studies with engineered nanoparticles. *Nanotoxicology*, *8*(3), 233–278.
35. Mocan, T., (2013). Hemolysis as expression of nanoparticles-induced cytotoxicity in red blood cells. *Biotechnol. Mol. Biol. Nanomedicine BMBN*, *1*, 7–12.
36. Schubert, M. A., & Müller-Goymann, C. C., (2005). Characterization of surface-modified solid lipid nanoparticles (SLN): Influence of lecithin and nonionic emulsifier. *European Journal of Pharmaceutics and Biopharmaceutics*, *61*(1), 77–86.
37. Manke, A., Wang, L., & Rojanasakul, Y., (2013). Mechanisms of nanoparticle induced oxidative stress and toxicity. *BioMed. Research International, 2013,* 1–15.
38. Syed, S., Zubair, A., & Frieri, M., (2013). Immune response to nanomaterials: Implications for medicine and literature review. *Current Allergy and Asthma Reports*, *13*(1), 50–57.
39. Pal, S. L., Jana, U., Manna, P. K., Mohanta, G. P., & Manavalan, R., (2011). Nanoparticle: An overview of preparation and characterization. *Journal of Applied Pharmaceutical Science, 1*(6), 228–234.

CHAPTER 3

Nano-Biomedicine: A Next-Generation Tool for Effective and Safe Therapy

VIKAS SHRIVASTAVA,* PALLAVI SINGH CHAUHAN, and RAJESH SINGH TOMAR

Amity Institute of Biotechnology, Amity University Madhya Pradesh, Gwalior – 474005, India

Corresponding author. E-mail: vshrivastava@gwa.amity.edu

ABSTRACT

Nanotechnology is an active area of research these days. Nanotechnology deals with the use of low energy and resources, small amount of nanomaterials are more effective as compared to large material in bulk amount. Nanotechnology is a combination of several fields like physics, chemistry, mathematics, material science, computer science, bio-instrumentation, environmental science, pharmacy, etc. These properties of nanoparticles make them more suitable to use in various biomedical and clinical applications such as drug delivery, tissue engineering, probing of DNA structure and antimicrobial agents.

3.1 INTRODUCTION

Nanoparticles (NPs) are known to have several interesting properties that make them useful to be used in several applications [1]. It is quite difficult to handle and synthesize NPs as compared to the bulk material [2]. The word nano is derived from the Greek word "nano," meaning dwarf [3]. One billionth of a meter is called nano, ranging in size less than 100 nm [4]. There are several bulk materials already reported to have more interesting properties and application area, when they are used in the nano range as it makes them useful to interact with other particles along with increasing their antibacterial potential [5]. The biological component as a synthesizing source of metal NPs is a method that has fewer hazards on the environment, easy to scale up, and enhanced stability [6].

3.1.1 NANOMEDICINE: CURRENT STATUS AND FUTURE PROSPECTS

According to the National Institutes of Health, Nanomedicine refers to applications of nanotechnology in the area of medicine like treatment, diagnosis, monitoring, and control of biological systems [7]. Drug delivery and drug targeting (involving identification of precise target) of pharmaceutical and therapeutic agents is the main target of using nanomedicine [8]. This way of delivery and targeting may minimize the side effects caused due to overdosing done for effective therapy [9].

There are a few key targets of nanomedicine like mononuclear phagocytes, dendritic cells, endothelial cells, and cancers [10]. Nowadays, nanotechnology is revolutionizing the industry, which includes particle designing and formulation [11]. Nanotechnology is providing a new advance in areas like disease diagnosis, its treatment, and prevention [12].

National Institutes of Health (Bethesda, MD, USA) gave these technologies and innovations a new term "nanomedicines" [13]. Biological mimetics, nanomachines, nanofibers, polymeric nanoconstructs, and nanoscale microfabrication-based devices are some of the sensors and laboratory diagnostics included in nanomedicine [14].

Furthermore, nanoscale materials are capable of targeting different cells along with delivering drugs to extracellular elements in the body [15]. Drug delivery and targeting via intravenous and interstitial routes of administration by means of NPs are advanced of projects in nanomedicine [16].

3.1.2 NANOTECHNOLOGIES FOR MEDICAL IMAGING AND TARGETED DRUG DELIVERY

Constructing of nanomaterials with unique properties is achieved by the manipulation of atoms and molecules [17]. Properties of an element can be improved by altering their size from their bulk form to nano [18]. NPs in the range of 1–20 nm are known to have vast applications in biology as well as in medicine [19]. In fluorescence resonance, energy transfer measurement studies gold nanoparticles (AuNPs) are used as quenchers [20]. In magnetic resonance imaging (MRI), Iron oxide nanocrystals as contrast agents are used, because of their changing ability in the spin-spin relaxation times of neighboring water molecules [21].

Labeling biological systems and their detection by means of optical or electrical means in vitro as well as *in vivo* by using quantum dots (QDs) is another example [22]. By alteration of particle size, the fluorescence emission wavelength of QDs can be tuned [23].

Recent reports have shown those metastatic tumor cell extravasations are tracked using QDs [24]. QDs are of prime concern just because multi-functional nanosystems can be designed due to their large area-to-volume ratio [25]. There is no doubt that such nanotools help us in providing true understanding about both normal as well as disease phases in humans.

3.1.3 NANOVEHICLES AND DRUG CARRIERS

The nanoscale is a unifying characteristic of any engineered construct like micelles, dendrimers, ceramics, polyplexes, and liposomes, just because of its unique properties when compared to the bulk one [26]. Encapsulation, covalent attachment as well as absorption of different therapeutic and diagnostic agents, can be done using nanocarriers, which ultimately reduces the risk of drug solubility, drug wastage along with different side effects [27]. Few compounds like dendrimers show less incorporation potential with active compounds [28]. Thus in this situation, milling of the smaller compound along with coating will be done so as to make them feasible for oral as well as intravenous delivery.

The reduced size of the particle thus provides faster as well as the quick release of drugs at a targeted site [29]. The NPs functioning, i.e., increased or reduced blood solubility, varies with respect to their pharmacokinetic profiles, the issue is to target them at a particular location with efficient drug release [30]. Thus, the use of NPs or nanocarriers may reduce the various risk associated with different biomedical applications like medical imaging, MRI contrast enhancement, oral delivery, targeted drug delivery, and reduced risks of side effects [31]. There are several factors affecting the engineering of nanocarriers along with their functioning, some of the factors discussed here are environmental pH, reducing agents, magnetic field, electric fields, or temperature [32].

The altered functioning of nanocarriers may lead to affect the various characteristics like the integrity of particles, delivery of drug, and the target for releasing drugs (targeted drug release) [33]. Multifunctionality is the prime focus of designing nanocarriers, and their multifunctionality includes the targeting of cell receptors, and drugs release along with developing sensors [34]. Encapsulation of QDs in micelle is an example showing the incorporation of more nanosystems within other carriers, and the entire process reduces the risk of quantum dotes nonspecific adsorption as well as aggregation [35].

The impact of such delivery systems can be seen in various pharmaceutical companies and their working process.

3.1.4 INNER SPACE

The chemical composition, structure as well as size of NPs affect their stability, distribution as well as behavior.

3.1.5 CLEARANCE MECHANISMS AND OPPORTUNITIES FOR TARGETING

Blood vessels and arteries are known to be the natural routes for the delivery of various nutrients, removal of toxic materials, and drug delivery [36]. But the controlled and specific access to various specific tissues and organs is somewhere lacking. The role of NPs is thus providing access to cells, tissues, and organs. Intravenous delivery provides rapid circulation followed by filtration using Kupffer cells and the marginal zone and red pulp [37]. The size and NPs surface play an essential role in the opsonization of blood and clearance as the bulk particles (200 nm and above) are more susceptible to efficiently activate the human complement system as compared to the NPs [38]. Surface property of NPs like types of functional groups; affect the binding process of blood proteins to opsonins.

3.1.6 TARGETING

3.1.6.1 MACROPHAGE AS A TARGET

Targeting macrophages with nanocarriers is provided by the propensity of macrophages with respect to recognition and particulate matter clearance [39]. The macrophage contribution to host defense mechanisms or pathogenesis is well known, but their dysfunctioning leads to several disorders such as atherosclerosis, autoimmunity, and major infections. For macrophage destruction, endocytic delivery is another way. For removing unwanted macrophages in various diseases like autoimmune blood disorders, rheumatoid arthritis, etc., nanocarrier mediated macrophage suicide is known to be a leading approach [40].

3.1.6.2 ENDOTHELIUM AS A TARGET

The endothelium has an important role in different pathological processes like cancer, inflammation, oxidative stress, and thrombosis. Various studies have reported for controlling the distribution of targeted NPs by specific endothelial cells [41]. Cationic liposomes after entering the circulation are internalized via endosomes and lysosomes.

3.1.6.3 EXTRAVASATION: TARGETING OF SOLID CANCERS

NPs are passively accumulated when injected called stealth technology [42], which is basically governed with drug delivery, imaging, etc. Perfusion heterogeneity leads to the generation of unpredictiveness of stealth NPs in solid tumors [43]. Liposomes with entrapped doxorubicin are few effective formulations have application in managing AIDS-related Kaposi's sarcoma etc. [44]. The most important aspect is that the loading capacity of the carrier must be a high drug with enhanced stability and reduced loss of drug [45].

3.1.7 NANOPARTICLES (NPS) FOR CYTOPLASMIC DRUG DELIVERY

Poly(DL-lactide-co-glycolide) synthesized NPs have the potential to escape the endo-lysosomal compartment and thus reaching to the cytoplasm, which becomes possible rapid by alteration of NPs surface charge from one state to another which leads to the local interaction with particle-membrane along with cytoplasmic content release [46]. Surface manipulation is another approach with short for cytoplasmic delivery.

3.1.8 TOXICITY ISSUES

Nanocarriers have enhanced drug solubility as well as stability property with less side effects. But the solution is always needed for the possibility of toxicity issues [47]. There are several reports which show the toxicity effects of nanotechnology, which are hazardous to the environment and biological systems. For example, human keratinocytes on exposure to carbon nanotubes (CNTs) may cause oxidative stress and apoptosis [48]. Intravenously, the injection of NPs showed enhanced toxicity [49].

3.1.9 CELL DEATH AND ALTERED GENE EXPRESSION

Recent studies have shown that cadmium selenide QDs have promising applications in imaging, but they release highly toxic cadmium ions, so they are lethal to cells under UV irradiation. Some studies have been reported for the polymeric micelles, inducing cell death via apoptosis or necrosis [50].

After cisplatin delivery using polymeric micelles, it has been reported that certain cells started showing altered gene expression as compared to the cells treated with the cells that are free of cisplatin treatment [51].

3.1.10 CELL DEATH AND GENE THERAPY

Viral vectors are known to be efficient delivery systems for nucleic acids, where they may lead to induce severe immunotoxicity when after randomly integrated into the host genome [52]. Thus the development of polycationic nonviral gene transfer systems is a prime concern, but they may also cause immediate or delayed cytotoxicity.

3.2 THE FUTURE OF NANOMEDICINE

Changes in the vascular imaging and drug delivery system are a new concern of nanotechnology. The field of nanomedicine may provide new initiatives that ultimately may lead to enhanced benefits, which basically involves the development of diagnostic devices along with drug discovery systems. The National Cancer Institute (NCI) is running several programs that are related to the disease diagnosis, deliver therapeutic agents, and monitor cancer treatment progress.

KEYWORDS

- **drug delivery**
- **drug targeting**
- **nanomedicine**
- **nanovehicles**
- **toxicity**

REFERENCES

1. Edina, C. W., & Andrew, Z. W., (2014). Nanoparticles and their applications in cell and molecular biology. *Integr. Biol., (Camb.), 6*(1), 9–26. doi: 10.1039/c3ib40165k.
2. Shiomi, S., Kawamori, M., Yagi, S., & Matsubara, E., (2015). One-pot synthesis of silica-coated copper nanoparticles with high chemical and thermal stability. *J. Colloid. Interface Sci., 460*, 47–54. doi: 10.1016/j.jcis.2015.08.033.
3. Suryyani, D., Kanjaksha, G., & Shrimati, D. S., (2015). Nanoimaging in cardiovascular diseases: Current state of the art. *Indian J. Med. Res. Mar., 141*(3), 285–298.
4. Albert, J. O., Juliane, N., Robert, B., & Quoc, T. D., (2015). Nanotechnology in respiratory medicine. *Respir. Res., 16*(1), 64. doi: 10.1186/s12931–015–0223–5.
5. Yi-Huang, H., Kuen-Song, L., Wan-Ju, K., Chien-Te, H., Chao-Lung, C., Dong-Ying, T., & Shih-Tung, L., (2015). The antimicrobial properties of silver nanoparticles in bacillus subtilis are mediated by released Ag+ ions. *PLoS One, 10*(12), e0144306. doi: 10.1371/journal.pone.0144306.
6. Ill-Min, C., Inmyoung, P., Kim, S. H., Muthu, T., & Govindasamy, R., (2016). Plant-mediated synthesis of silver nanoparticles: Their characteristic properties and therapeutic applications. *Nanoscale Res. Lett., 11*, 40. doi: 10.1186/s11671–016–1257–4.
7. Pothur, R. S., Martin, P., Tania, Q. V., Qingrong, H., Josef, L. K., Etta, S., et al., (2010). Nanotechnology research: Applications in nutritional sciences. *J. Nutr. Jan., 140*(1), 119–124. doi: 10.3945/jn.109.115048.
8. Sadat, S. M., Saeidnia, S., Nazarali, A. J., & Haddadi, A., (2015). Nano-pharmaceutical formulations for targeted drug delivery against HER2 in breast cancer. *Curr. Cancer Drug Targets, 15*(1), 71–86.
9. Charu, B., Upendra, N., Ashok, K. P., & Neha, G., (2015). Mesoporous silica nanoparticles in target drug delivery system: A review. *Int. J. Pharm. Investig., 5*(3), 124–133. doi: 10.4103/2230–973X.160844.
10. Aldrin, E. S., (2015). Nanomedicine concepts in the general medical curriculum: Initiating a discussion. *Int. J. Nanomedicine, 10*, 7319–7331. doi: 10.2147/IJN.S96480.

11. Ernie, H., (2004). Nanotechnology: Looking as we leap. *Environ. Health Perspect., 112*(13), A740–A749.
12. Ahmed, S., Basma, M. E., Hala, G., & Hatem, E. S., (2015). Nanotechnology applications in hematological malignancies (Review). *Oncol. Rep., 34*(3), 1097–1105. doi: 10.3892/or.2015.4100.
13. Jangsun, H., Yoon, J., Jeong, M. P., Kwan, H. L., Jong, W. H., & Jonghoon, C., (2015). Biomimetics: Forecasting the future of science, engineering, and medicine. *Int. J. Nanomedicine, 10*, 5701–5713. doi: 10.2147/IJN.S83642.
14. Jinjun, S., Alexander, R. V., Omid, C. F., & Robert, L., (2010). Nanotechnology in drug delivery and tissue engineering: From discovery to applications. *Nano Lett., 8, 10*(9), 3223–3230. doi: 10.1021/nl102184c.
15. Jiahe, L., Charles, C. S., Dantong, H., & Michael, R. K., (2015). Nanobiotechnology for the therapeutic targeting of cancer cells in blood. *Cell Mol. Bioeng., 8*(1), 137–150. doi: 10.1007/s12195–015–0381-z.
16. David, S. H., Aniket, S. W., Nathan, B. R., Jimena, G. P., Nina, P. C., Victor, F., Jeffrey, A. W., Graeme, F. W., & Anthony, J. K., (2016). Evolving drug delivery strategies to overcome the blood brain barrier. *Curr. Med. Chem., 22*(9), 1177–1193. doi: 10.2174/1381612822666151221150733.
17. Salata, O. V., (2004). Applications of nanoparticles in biology and medicine. *J. Nanobiotechnology, 2*, 3. doi: 10.1186/1477–3155–2–3.
18. Yvoni, K., Margarita, S., Maria-Eleni, D., Vincent, B., Athina, B., Alexander, T., Nikolaos, M., & Konstantinos, M., (2016). New Ti-alloys and surface modifications to improve the mechanical properties and the biological response to orthopedic and dental implants: A review. *Biomed. Res. Int.*, 2908570. doi: 10.1155/2016/2908570.
19. Erik, C. D., Lauren, A. A., Megan, A. M., & Mostafa, A. E. S., (2012). Size matters: Gold nanoparticles in targeted cancer drug delivery. *Ther. Deliv. Ther. Deliv., 3*(4), 457–478.
20. Chen, J., Huang, Y., Zhao, S., Lu, X., & Tian, J., (2012). Gold nanoparticles-based fluorescence resonance energy transfer for competitive immunoassay of biomolecules. *Analyst, 137*(24), 5885–5890. doi: 10.1039/c2an36108f.
21. Wenbin, L., Taeghwan, H., Gregory, M. L., Miqin, Z., & Thomas, J. M., (2009). Magnetic nanoparticles for early detection of cancer by magnetic resonance imaging. *MRS Bull., 34*(6), 441–448. doi: 10.1557/mrs2009.120.
22. Mei-Xia, Z., & Er-Zao, Z., (2015). Application of functional quantum dot nanoparticles as fluorescence probes in cell labeling and tumor diagnostic imaging. *Nanoscale Res. Lett., 10*, 171. doi: 10.1186/s11671–015–0873–8.
23. Andrew, M. S., & Shuming, N., (2010). Semiconductor nanocrystals: Structure, properties, and band gap engineering. *Acc. Chem. Res., 16, 43*(2), 190–200. doi: 10.1021/ar9001069.
24. Nobuyuki, K., Thomas, E. M., Makoto, M., Peter, L. C., & Hisataka, K., (2010). Real-time optical imaging using quantum dot and related nanocrystals. *Nanomedicine (Lond), 5*(5), 765–776. doi: 10.2217/nnm.10.49.
25. Luke, J. M., Renea, F., Supriya, R., Hong, Z., & Lisa, A., (2015). Delouise. UVB dependence of quantum dot reactive oxygen species generation in common skin cell models. *J. Biomed. Nanotechnol., 11*(9), 1644–1652.

26. Iseult, L., Ilise, L. F., & Michaela, K., (2015). 'Bio-nano interactions: New tools, insights and impacts': Summary of the Royal Society discussion meeting. *Philos. Trans. R. Soc. Lond. B. Biol. Sci., 370*(1661), 20140162. doi: 10.1098/rstb.2014.0162.
27. Paresh, C. R., Hongtao, Y., & Peter, P. F., (2009). Toxicity and environmental risks of nanomaterials: Challenges and future needs. *J. Environ. Sci. Health C. Environ. Carcinog. Ecotoxicol. Rev., 27*(1), 1–35. doi: 10.1080/10590500802708267.
28. Tristan, B., Gregory, R. M. D., Peter, L., Choyke, M. D., & Hisataka, K., (2009). Dendrimers application related to bioimaging. *IEEE Eng. Med. Biol. Mag., 28*(1), 12–22. doi: 10.1109/MEMB.2008.931012.
29. Garima, S., Ashish, R. S., Ju-Suk, N., George, P., Doss, C., Sang-Soo, L., & Chiranjib, C., (2015). Nanoparticle based insulin delivery system: The next generation efficient therapy for type 1 diabetes. *J. Nanobiotechnology, 13*, 74. doi: 10.1186/s12951–015–0136-y.
30. Mangoni, A. A., & Jackson, S. H. D., (2004). Age-related changes in pharmacokinetics and pharmacodynamics: Basic principles and practical applications. *Br. J. Clin. Pharmacol., 57*(1), 6–14. doi: 10.1046/j.1365–2125.2003.02007.x.
31. Hui, Y. X., Pengbo, G., Wu-Cheng, W., & Ho, L. W., (2015). Lipid-based nanocarriers for RNA Delivery. *Curr. Pharm. Des., 21*(22), 3140–3147. doi: 10.2174/138161282 1666150531164540.
32. Rajendran, J. C., Bose, S. H. L., & Hansoo, P., (2016). Lipid-based surface engineering of PLGA nanoparticles for drug and gene delivery applications. *Biomater. Res., 20*, 34. doi: 10.1186/s40824–016–0081–3.
33. Rajesh, S., & James, W. L., (2009). Nanoparticle-based targeted drug delivery. *Jr. Exp. Mol. Pathol., 86*(3), 215–223. doi: 10.1016/j.yexmp.2008.12.004.
34. Micah, D., Glasgow, K., & Mahavir, B. C., (2015). Recent developments in active tumor targeted multifunctional nanoparticles for combination chemotherapy in cancer treatment and imaging. *J. Biomed. Nanotechnol., 11*(11), 1859–1898.
35. Lemon, C. M., Karnas, E., Han, X., Bruns, O. T., Kempa, T. J., Fukumura, D., Bawendi, M. G., Jain, R. K., Duda, D. G., & Nocera, D. G., (2015). Micelle-encapsulated quantum dot-porphyrin assemblies as in vivo two-photon oxygen sensors. *J. Am. Chem. Soc., 137*(31), 9832–9842. doi: 10.1021/jacs.5b04765.
36. *Blood Vessels and Endothelial Cells* (4th edn.). Molecular Biology of the Cell.
37. Bonnie, H., Wuhbet, D. A., Yiran, Z., Sandra, C., Bustamante, L., Samantha, S. L., & Darrell, J. I., (2015). Active targeting of chemotherapy to disseminated tumors using nanoparticle-carrying T cells. *Sci. Transl. Med., 7*(291), 291ra94. doi: 10.1126/scitranslmed.aaa5447.
38. Heather, H. G., Dolly, H. C., David, W. G., & Hamidreza, G., (2015). Nanoparticle uptake: The phagocyte problem. *Nano Today, 10*(4), 487–510. doi: 10.1016/j.nantod.2015.06.006.
39. Bagalkot, V., Badgeley, M. A., Kampfrath, T., Deiuliis, J. A., Rajagopalan, S., & Maiseyeu, A., (2015). Hybrid nanoparticles improve targeting to inflammatory macrophages through phagocytic signals. *J. Control Release, 217*, 243–255. doi: 10.1016/j.jconrel.2015.09.027.
40. Saxena, T., Loomis, K. H., Pai, S. B., Karumbaiah, L., Gaupp, E., Patil, K., Patkar, R., & Bellamkonda, R. V., (2015). Nanocarrier-mediated inhibition of macrophage

migration inhibitory factor attenuates secondary injury after spinal cord injury. *ACS Nano., 9*(2), 1492–1505. doi: 10.1021/nn505980z.Epub.
41. Rashmi, H. P., Vandana, B. P., & Medha, D. J., (2015). Polymeric nanoparticles for targeted treatment in oncology: current insights. *Int. J. Nanomedicine, 10*, 1001–1018. doi: 10.2147/IJN.S56932.
42. Gopalakrishna, P., & Maria, L. C. C., (2013). Science and technology of the emerging nanomedicines in cancer therapy: A primer for physicians and pharmacists. *SAGE Open Med., 1,* 2050312113513759. doi: 10.1177/2050312113513759.
43. Lisa, S., Tejaswi, V., Fatemeh, M., Sherry, Y. W., Anil, K. S., & Susan, H., (2015). Advances and challenges of liposome assisted drug delivery. *Front Pharmacol., 6*, 286. doi: 10.3389/fphar.2015.00286.
44. Lee, R. J., & Low, P. S., (1995). Folate-mediated tumor cell targeting of liposome-entrapped doxorubicin *in vitro*. *Biochim. Biophys. Acta., 1233*(2), 134–144.
45. Meysam, M., Kambiz, G., & Seyed, A. M., (2015). Preparation and characterization of Rifampin loaded mesoporous silica nanoparticles as a potential system for pulmonary drug delivery. *Iran J. Pharm. Res. Winter, 14*(1), 27–34.
46. Gomes, C., Moreira, R. G., & Castell-Perez, E., (2011). Poly (DL-lactide-co-glycolide) (PLGA) nanoparticles with entrapped trans-cinnamaldehyde and eugenol for antimicrobial delivery applications. *J. Food Sci., 76*(2), 16–24. doi: 10.1111/j.1750–3841.2010.01985.x.
47. Mayank, S., Rajesh, S., & Dinesh, K. J., (2016). Nanotechnology based approaches for enhancing oral bioavailability of poorly water soluble antihypertensive drugs. *Scientifica (Cairo).,* 8525679. doi: 10.1155/2016/8525679.
48. Katsuhide, F., Makiko, F., Shigehisa, E., Junko, M., Haruhisa, K., Ayako, N., Naohide, S., Kanako, U., & Kazumasa, H., (2015). Size effects of single-walled carbon nanotubes on *in vivo* and *in vitro* pulmonary toxicity. *Inhal. Toxicol., 27*(4), 207–223. doi: 10.3109/08958378.2015.1026620.
49. Jong-Suep, B., Ju-Heon, K., Jeong-Sook, P., & Cheong-Weon, C., (2015). Modification of paclitaxel-loaded solid lipid nanoparticles with 2-hydroxypropyl-β-cyclodextrin enhances absorption and reduces nephrotoxicity associated with intravenous injection. *Int. J. Nanomedicine, 10*, 5397–5405. doi: 10.2147/IJN.S86474.
50. Aditya, K. J., Sahabjada, S. M. L., Irfan, A. K., & Md. Arshad, (2016). Chemically synthesized CdSe quantum dots inhibit growth of human lung carcinoma cells via ROS generation. *EXCLI J., 15*, 54–63. doi: 10.17179/excli2015–705.
51. Xiang, K., Hai-Hua, X., Hai-Qin, S., Xia-Bin, J., Le-San, Y., & Ruo-Gu, Q., (2015). Advances in drug delivery system for platinum agents based combination therapy. *Cancer Biol. Med., 12*(4), 362–374. doi: 10.7497/j.issn.2095–3941.2015.0063.
52. Nouri, N., Talebi, M., & Palizban, A. A., (2012). Viral and nonviral delivery systems for gene delivery. *Adv. Biomed Res., 1*, 27. doi: 10.4103/2277–9175.98152.

CHAPTER 4

Nanotheranostics: Next Generation Diagnosis and Therapy for Cancer

SHARMISTHA BANERJEE,[1] SHUCHI KAUSHIK,[2] and RAJESH SINGH TOMAR[1]

[1]*Amity Institute of Biotechnology, Amity University Madhya Pradesh, Gwalior – 474005, India*

[2]*M.P. Forensic Science Laboratory, Madhya Pradesh, India*

[*]*Corresponding author. E-mail: sbanerjee@gwa.amity.edu (S. Banerjee); skaushik@gwa.amity.edu (S. Kaushik); rstomar@amity.edu (R. S. Tomar)*

ABSTRACT

The term "theranostics" is used to develop more specific and personalized treatment approaches for various diseases by combining diagnostic and therapeutic modality into a single regimen. The concept was developed from the fact that disease, such as cancer, is extremely heterogeneous, and all existing treatment methods are effective for only a specific subpopulation of patients and that too at selective stages of disease progression.

4.1 INTRODUCTION

Cancer is one of the major menaces to human health in the present scenario [1]. Early diagnosis and effective treatment approaches to this disease may help to overcome this situation. It has been noted that the prime reason behind death due to cancer is tumor metastasis. Since cancer cells circulate through the bloodstream, researchers are working to develop new approaches to detect circulating tumor cells (CTCs) in early stages to help doctors treat patients and predict the development of cancer. Even though CTCs were discovered long back in 1869, these can be identified in patients bearing tumors only at later stages. Because of their extremely small quantity that accounts for only a few cells/10^6 peripheral blood mononuclear cells, its early identification is, therefore, a challenging task. Recent researches show that using nanomaterials as contrast agents and therapeutic actuators will prove to be a potential treatment approach, though it is still at its initial stage. Since nanoparticles (NPs) pose unique magnetic, photothermal, and optical properties, they can be used not only for selective identification but also as non-invasive photothermal therapy for treating cancer [2].

During the last five years, research shows that diagnosis and therapy have to be combined in a single regimen nanomaterial, known as "theranostic NPs." The ideal theranostic nanomaterials possess the following advantages: (1) it has the ability to selectively accumulate in the diseased tissue, (2) capacity to deliver an effective therapeutic action selectively, and (3) is safe and has the ability to be biodegraded into nontoxic by-products [2].

Nanotechnology has been able to bring therapy and diagnosis close to one another. Nanoparticle (NP)-based imaging and therapeutic approach have been explored individually, and these researches have led to the emergence of NP-based theranostics, which are basically nanoplatforms that can perform imaging and therapeutic actions simultaneously. This is an advancement over the traditional theranostics. It also provides an added advantage over the previous patterns that not only allows images to be captured before or after the treatment approaches but also simultaneously during the course of therapy. These are suitable because several nanomaterials that are already being used as imaging agents can be easily "upgraded" to theranostic agents by associating therapeutic compounds with them [3].

NPs have been combined with various imaging agents, targeting moieties and cargo drugs, thus generating theranostic NPs that have the capacity to deliver therapeutic agents with the simultaneous diagnosis. Since, last ten years, it was observed that NPs in size range of 4–100 nm that is almost 1000–10,000 times smaller in comparison to healthy cells of humans, show an intense relationship with biomolecules like receptors, antibodies, and enzymes, both on the interior and the exterior surface of the cell. NPs can be functionalized by coating and chemical modification and also be integrated with several biological moieties for selective diagnosis and therapy. Thus, it is believed that nanomedicines will bring a revolution in diagnosis, detection, and cancer therapy [2].

4.2 THERANOSTIC NANOMEDICINES

Theranostic nanomedicines can be categorized into two main types:

1. Inorganic; and
2. Organic.

4.2.1 INORGANIC

Inorganic materials like iron oxide, silica, gold, and various other metals are used for the preparation of theranostic NPs [4].

4.2.1.1 GOLD NANOPARTICLES (AUNPS)

CTC can be detected by a multifunctional plasmonic shell-magnetic core nanotechnology-driven approach for specific diagnosis and cancer cell isolation [5, 6].

Nanomaterials, such as gold nanoparticles (AuNPs), show unique properties from their bulk counterparts. Due to the spatial confinement of the conduction electrons, distinct optical characteristics emerge [7–10]. The intensity and frequency at which the localized surface plasmon resonance (LSPR) arises can be tuned by varying metal composition, size, shape, and dielectric medium [11, 12]. The light extinction generated by the LSPR includes both scattering and absorption phenomena [13]. The scattering element of the extinction band is directly proportional to the size of the particle, whereas absorbance is inversely related to the size of the particle. AuNPs' spectral behavior can be exploited in many different spectral regions. When AuNP absorbs light, it decomposes in non-radiative mode to a level of low energy with heat production. This property can be used in treatment approaches like photothermal therapy. LSPR is not only tunable, but its extinction coefficient for laser absorption is five times greater in comparison to that of organic dyes, which were conventionally used in photothermal therapy. AuNPs are also known to show resistance towards photobleaching [9].

AuNPs are exceptionally versatile as they can be used as an imaging probe or can be easily functionalized with appropriate imaging agents, which would overcome some of the limitations of presently existing techniques [14].

Tumor thermal therapy using AuNPs is one of the most explored research areas in the 2000s [15]. But AuNPs alone show low targeting ability and are relatively less stable in blood. Modified AuNPs, for example, polyethylene glycol (PEG) – Transferrin (Tf) grafted AuNPs, showed enhanced uptake by cells and selective delivery to tumors specifically with lesser toxicity [4].

Apart from AuNPs, mesoporous silica nanoparticle is an alternative method for the selective delivery of drugs [2].

4.2.1.2 MESOPOROUS SILICA

Presently, mesoporous silica nanoparticles (MSNs) have been observed as prospective candidates for theranostic, mainly because of their tailorable size and diameter of the pore, outstanding biocompatibility, intrinsically increased surface area, and topologically different regions that can be functionalized independently [16–22]. On the basis of these unique properties, numerous scientists have attempted to expand the usefulness of MSNs further by the creation of composite inorganic/organic hybrid MSNs [23–26]. The inherent biocompatibility of MSNs and the simplicity by which they can be modified chemically using both covalent and ionic ligands after synthesis has led to the development of numerous inorganic/organic/MSN conjugations. MSNs have three topologically different domains that can be functionalized individually: hexagonal nanochannels/pores, particle exterior, and the silica framework. MSNs are especially a suitable candidate for theranostic applications; with separate domains that can be used as (1) contrast agents in tracing the image of targeted diagnosis (2) biomolecular ligands, for specific delivery of both conveyed cargo and drug platform and (3) payloads, for therapeutic intervention [27].

In order to improve the bio-functionality of MSNs, significant efforts have been put for developing organic hybrid MSNs: MSNs that include organic polymers like polyethylene, polyethylene glycol, or pH-sensitive polymers on their interior or exterior surfaces. Organic hybridization of MSNs can make them better agents to react to changes in their local surroundings (i.e., switchable "smart" nanomaterials). But organic hybridization of MSNs has led to its increasing use for stabilizing payloads for the conveyance or increasing the possibility of efficient targeting. For example, the PEGylation of NPs usually enhances its circulation time, provides resistance to proteolysis, and decreases their immunogenicity and/or antigenicity [27].

4.2.1.3 IRON-OXIDE NANOPARTICLES (NPS)

Iron oxide nanoparticles (IONPs) are nanocrystals that are made of hematite or magnetite. Much different from their bulk counterparts, IONPs that are lesser than 20 nm are superparamagnetic--a condition where particles show zero magnetism in the absence of an external magnetic field, but maybe magnetized in the presence of an external magnetic field. Better magnetic properties of IONPs, along with their inherent inexpensiveness and biocompatibility, have made them a potential candidate to be used in several biomedical applications like contrast probes for magnetic resonance imaging (MRI). IONPs with suitable coatings can be readily associated with therapeutic agents. For example, Zhang and his group coupled an anti-cancer drug (methotrexate), with IONP functionalized by amines. *In vitro* studies shows that the NPs, after being internalized by cells, get accumulated in lysosomes, where the drugs are released because of the presence of proteases and comparatively lesser pH. Hwu et al. in 2009 [28] reported the association of paclitaxel (PTX) to IONP surfaces through a phosphodiester group at the ($Carbon_2$)-OH position. The average number of PTX molecules coupled per nanoparticle was found as 83, and the drug release was observed to be highly efficient in the presence of phosphodiesterase [3].

IONP can be used as both imaging or therapy because of its potent action in hyperthermia. The technique involved is IONPs can act as antennae in the presence of an external alternating magnetic field (AMF) that could change electromagnetic energy to heat energy. This property can be exploited in treating tumor cells that are highly sensitive to increased temperature in comparison to healthy cells. In an instance, phospholipid coated IONPs were administered subcutaneously in F344 model rats bearing tumor and were kept in the presence of an AMF. The AMF in association with IONPs elevated the tumor cell's temperature greater than 43°C resulting in regression of the tumor, but it does not show any effect on the control group (unexposed to IONP). Also, the Fab fragment of anti-human MN antigen-specific antibody was coupled chemically on the surface of IONP and was injected through a systemic route into mice suffering from a tumor. These particles exhibited efficient uptake of tumor cells, possibly because of antigen-antibody interaction, and induced tumor hyperthermia in the presence of an AMF [3].

4.2.2 ORGANIC NANOPARTICLES (NPS)

4.2.2.1 LIPID-BASED NANOPARTICLES (NPS)

Because of minimum toxicity and ready uptake by cells, lipid-based NPs play an important role in cancer therapy. Due to the Enhanced Permeability and Retention effect and longer half-life in lipids, liposomes, and blood, they are widely used as basic components for creating theranostic NPs since conventional times of nanobiotechnology [29, 30]. Lipid nanosuspensions, the very first concept of lipid-based nanoformulation, were used in clinics for the supply of nutrients in the 1950s [31]. Most popularly utilized lipid-based materials for NP formulation like phosphatidylcholine, 1,2-distearoyl-*sn*-glycero-3-phosphoethanolamine-*N*-methoxypoly(ethylene glycol) 2000 and cholesterol, are all recommended by the US FDA (United States Food and Drug Administration) [32, 33]. Lipid-based nanocarriers include liposomes, lipid micelles, solid-lipid NPs, nanosuspensions, and nanoemulsions. Particularly, liposomes have been studied as theranostic NPs for cancer for the past 30 years [34].

Liposomes are colloid based self-closed vesicles that form a hydrophobic patch in between the inner core and lipid bilayer membrane. Liposomal NPs can be categorized into small unilamellar (approximately 100 nm in diameter), large unilamellar (200–800 nm in diameter), multilamellar (500–5,000 nm in diameter) and liposomes according to the number of layers present in it [35]. Liposomal NPs are biocompatible vesicles that are loaded with either hydrophobic or hydrophilic therapeutic agents in their multimeric bilayers. Practically, numerous hydrophobic chemodrugs for cancer, such as doxorubicin and paclitaxel, have been loaded in liposomal vesicles of nano range, and they are already present in the market or in clinical phase [36, 37].

4.2.2.2 POLYMER-BASED NANOPARTICLES (NPS)

Polymers are the most suitable base materials for creating the base of NPs. In a previous report of 1979 on the use of a polymer for treatment of cancer, the development of polymeric nanomedicine was fostered because of utmost necessity for newer options of drug delivery with improved characteristics like easily tunable physical and chemical properties, sustained release of anti-cancer drugs, biodegradability, biocompatibility, etc. [29]. A large number of polymers have been used to create theranostic NPs, which can be categorized into two major groups: natural and synthetic polymers. Natural polymers include proteins and polysaccharides, but polysaccharides are the most popularly used natural polymers for the synthesis of NPs [4].

Polysaccharides have several functional groups in their molecular chains that are responsible for easy modification in fabricating the structure of NPs. In addition, hydrophilic groups like carboxyl, amino, and hydroxyl functional groups are necessary for their bio-adhesive properties and their ability to form non-covalent bonds with glycoproteins present on the surface of mucosal cells [38]. Among polysaccharide-based NPs, chitosan is the most commonly used since the conventional times of theranostic NPs. Chitosan, which is a deacetylated chitin, is a natural polymer consisting of β-(1,4)-2-amido-D-glucose linked through (1-4)-glycosidic bonds. A wide range of chemically modified chitosan derivatives that includes trimethyl chitosan, carboxymethyl chitosan, glycol chitosan, or PEGylated chitosan has been commonly used in cancer theranostics [39–42].

Hyaluronic acid, which is a negatively charged copolymer of *N*-acetyl-D-glucosamine and D-glucuronic acid, is also the most commonly used polysaccharide for creating NPs [43, 44]. As the HA receptor (CD44) is overexpressed in most of the cancer cells, HA is used as an anti-cancer macromolecule due to the increased affinity of hyaluronic acid to CD44 in cancer cells [45–47]. Presently, PEGylated hyaluronic acid NPs and self-assembled hyaluronic acid NPs, which consisted of amphiphilic Hyaluronic acid -5-β-cholanic acid conjugates were prepared and characterized for cancer theranostic agents [43, 44]. The PEGylated hyaluronic acid NPs showed efficient delivery of drugs in tumors. These particles could release the loaded anticancer agents in the presence of hyaluronidase-1, an enzyme that is known to be responsible for the proliferation of cancer cells and neovascularization of tumor cells. This enzyme-specific drug release allows the delivery of the drug specifically to the tumor cells. HA-based NPs were utilized in optical and ultrasound imaging applications [4].

Perfluoropentane is a hydrophobic molecule that has a boiling point of 27° C in its gaseous phase at 37°C. Hence, it is utilized as a contrast probe in ultrasound imaging. Perfluoropentane was encapsulated in hyaluronic acid NPs to differentiate a tumor cell from healthy cells by exhibited showed surprisingly increased echogenicity [48]. Other polysaccharides such as heparin, alginate, dextran, and cellulose can also be utilized as a base material for creating the structure of NPs [43, 49].

Heparin is a sulfated repeating disaccharide unit, which also serves as a common base material to build NPs because of its potential to efficiently deliver protein/peptide drugs specifically to tumor cells. Increased affinity of dextran derivatives for iron oxide makes dextran NPs a suitable candidate as a MRI nanoprobe. Iron oxide associated with dextran NPs has drawn surveillance for clinical imaging of cancer cells using MRI [4].

PEG is one of the other most commonly utilized synthetic polymers. With the first approval of PEG for clinical use in the 1990s, it is considered as the gold standard of drug carriers amongst all other polymers. It is used popularly due to its solubility, hydrophilicity, and also because of its minimal interference with blood components and the immune system. Other benefits of PEG include lesser hydrodynamic volume, low PDI, increased solubility in organic solvents, etc. [4].

The very first agent that was discovered as both targeting moiety as well as therapeutic molecule was Herceptin®. It is a humanized antibody that has the capacity of targeting and blocking the overexpression of HER2 protein. It is used for treating metastatic breast cancers, which are HER2 positive [14].

KEYWORDS

- **circulating tumor cells**
- **gold nanoparticles**
- **localized surface plasmon resonance**
- **mesoporous silica nanoparticles**
- **nanoparticle**
- **polyethylene glycol**

REFERENCES

1. Chen, Q., Ke, H., Dai, Z., & Liu, Z., (2015). Nanoscale theranostics for physical stimulus-responsive cancer therapies. *Biomaterials, 73*, 214–230.
2. Fan, Z., Fu, P. P., Yu, H., & Ray, P. C., (2014). Theranostic nanomedicine for cancer detection and treatment. *Journal of Food and Drug Analysis, 22*, 3–17.

3. Xie, J., Lee, S., & Chen, X., (2010). Nanoparticle-based theranostic agents. *Adv. Drug Deliv. Rev., 62*(11), 1064–1079.
4. Yhee, J. Y., Son, S., Kim, N., Choi, K., & Kwon, I. C., (2014). Theranostic applications of organic nanoparticles for cancer treatment. *MRS Bulletin, 39*, 239–249.
5. Fan, Z., Shelton, M., Singh, A. K., et al., (2012a). Multifunctional plasmonic shellemagnetic core nanoparticles for targeted diagnostics, isolation, and photothermal destruction of tumor cells. *ACS Nano, 6*, 1065–1073.
6. Fan, Z., Senapati, D., Singh, A. K., et al., (2012b). Theranostic magnetic core-plasmonic shell star shape nanoparticle for the isolation of targeted rare tumor cells from whole blood, fluorescence imaging, and photothermal destruction of cancer. *Mol. Pharm., 10*, 857–866.
7. Tabor, C., Murali, R., Mahmoud, M., & El-Sayed, M. A., (2009). On the use of plasmonic nanoparticle pairs as a plasmon ruler: The dependence of the near-field dipole plasmon coupling on nanoparticle size and shape. *J. Phys. Chem., A., 113*(10), 1946–1953.
8. Pelton, M., Aizpurua, J., & Bryant, G., (2008). Metal-nanoparticle plasmonics. *Laser Photonics Rev., 2*(3), 136–159.
9. Jain, P. K., Lee, K. S., El-Sayed, I. H., & El-Sayed, M., (2006). Calculated absorption and scattering properties of gold nanoparticles of different size, shape, and composition: Applications in biological imaging and biomedicine. *J. Phys. Chem. B., 110*(14), 7238–7248.
10. Ghosh, S. K., & Pal, T., (2007). Interparticle coupling effect on the surface plasmon resonance of gold nanoparticles: From theory to applications. *Chem. Rev., 107*(11), 4797–4862.
11. Hu, M., Chen, J., & Li, Z. Y., (2006). Gold nanostructures: Engineering their plasmonic properties for biomedical applications. *Chem. Soc Rev., 35*(11), 1084–1094.
12. Lin, A. Y., Young, J. K., Nixon, A. V., & Drezek, R. A., (2014). Synthesis of a quantum nanocrystal-gold nanoshell complex for near-infrared generated fluorescence and photothermal decay of luminescence. *Nanoscale, 6*(18), 10701–10709.
13. Link, S., & El-Sayed, M., (2000). Shape and size dependence of radiative, non-radiative and photothermal properties of gold nanocrystals. *Int. Rev. Phys. Chem., 19*(3), 409–453.
14. Vinhas, R., Cordeiro, M., Carlos, F. F., Mendo, S., Fernandes, A. R., Figueiredo, S., & Baptista, P. V., (2015). Gold nanoparticle-based theranostics: Disease diagnostics and treatment using a single nanomaterial. *Nanobiosensors in Disease Diagnosis, 4*, 11–23.
15. Visaria, R. K., Griffin, R. J., Williams, B. W., Ebbini, E. S., Paciotti, G. F., Song, C. W., & Bischof, J. C., (2006). Enhancement of tumor thermal therapy using gold nanoparticle-assisted tumor necrosis factor-alpha delivery. *Mol. Cancer Ther., 5*(4), 1014–1020.
16. Lee, J. E., Lee, N., Kim, T., Kim, J., & Hyeon, T., (2011). Multifunctional mesoporous silica nanocomposite nanoparticles for theranostic applications. *Acc. Chem. Res., 44*(10), 893–902.
17. Li, Z., Barnes, J. C., Bosoy, A., Stoddart, J. F., & Zink, J. I., (2012). Mesoporous silica nanoparticles in biomedical applications. *Chem. Soc. Rev., 41*(7), 2590–2605.
18. Mamaeva, V., Sahlgren, C., & Linden, M., (2012). *Adv. Drug Delivery Rev.*, 00248.

19. Yang, Y. W., (2011). Towards biocompatible nanovalves based on mesoporous silica nanoparticles. *Med. Chem. Commun., 2*(11), 1033.
20. Wu, S. H., Hung, Y., & Mou, C. Y., (2011). Mesoporous silica nanoparticles as nanocarriers. *Chem. Commun., 47*(36), 9972–9985.
21. Tang, F., Li, L., & Chen, D., (2012). Mesoporous silica nanoparticles: Synthesis, biocompatibility and drug delivery. *Adv. Mater., 24*(12), 1504–1534.
22. Huang, W. Y., & Davis, J. J., (2011). Multimodality and nanoparticles in medical imaging. *Dalton Trans., 40*(23), 6087–6103.
23. Hu, C. M., Aryal, S., & Zhang, L., (2010). Nanoparticle-assisted combination therapies for effective cancer treatment. *Ther. Delivery, 1*(2), 323–334.
24. Kim, T. W., Slowing, I. I., Chung, P. W., & Lin, V. S., (2011). Ordered mesoporous polymer-silica hybrid nanoparticles as vehicles for the intracellular controlled release of macromolecules. *ACS Nano, 5*(1), 360–366.
25. Wang, L. S., Wu, L. C., Lu, S. Y., Chang, L. L., Teng, I. T., Yang, C. M., & Ho, J. A., (2010). Biofunctionalized phospholipid-capped mesoporous silica nanoshuttles for targeted drug delivery: Improved water suspensibility and decreased nonspecific protein binding. *ACS Nano, 4*(8), 4371–4379.
26. Baeza, A., Guisasola, E., Hernandez, E. R., & Regı, M. V., (2012). Magnetically triggered multidrug release by hybrid mesoporous silica nanoparticles. *Chem. Mater., 24*(3), 517–524.
27. Chen, N. T., Cheng, S. H., Souris, J. S., Chen, C. T., Moub, C. Y., & Lo, L. W., (2013). Theranostic applications of mesoporous silica nanoparticles and their organic/inorganic hybrids. *Journal of Materials Chemistry B, 1*, 3128–3135.
28. Hwu, J. R., Lin, Y. S., Josephrajan, T., Hsu, M. H., Cheng, F. Y., Yeh, C. S., Su, W. C., & Shieh, D. B., (2009). Targeted Paclitaxel by conjugation to iron oxide and gold nanoparticles. *J. Am. Chem. Soc., 131*(1), 66–68.
29. Peer, D., Karp, J. M., Hong, S., Farokhzad, O. C., Margalit, R., & Langer, R., (2007). Nanocarriers as an emerging platform for cancer therapy. *Nat. Nanotechnol., 2*(1), 751–760.
30. Mendoza, A. E. H., Campanero, M. A., Mollinedo, F., & Prieto, M. J., (2009). Lipid nanomedicine for anticancer drug therapy. *J. Biomed. Nanotechnol., 5*(4), 323–343.
31. Altmayer, P., Grundmann, U., Ziehmer, M., & Larsen, R., (1993). *AINS, 28*, 415.
32. Koo, O. M., Rubinstein, I., & Onyuksel, H., (2005). Role of nanotechnology in targeted drug delivery and imaging: A concise review. *Nanomedicine, 1*(3), 193–212.
33. Barenholz, Y., (2012). Doxil®--the first FDA-approved nano-drug: Lessons learned. *J. Control. Release, 160*(2), 117–134.
34. Malam, Y., Loizidou, M., & Seifalian, A. M., (2009). Liposomes and nanoparticles: Nanosized vehicles for drug delivery in cancer. *Trends Pharmacol. Sci., 30*(11), 592–599.
35. Torchilin, V. P., (2005). Recent advances with liposomes as pharmaceutical carriers. *Nat. Rev. Drug Discov., 4*(2), 145–160.
36. Constantinides, P. P., Chaubal, M. V., & Shorr, R., (2008). Advances in lipid nanodispersions for parenteral drug delivery and targeting. *Adv. Drug Deliv. Rev., 60*(6), 757–767.
37. Samad, A., Sultana, Y., & Aqil, M., (2007). Liposomal drug delivery systems: An update review. *Curr. Drug Deliv., 4*(4), 297–305.

38. Liu, Z. H., Jiao, Y. P., Wang, Y. F., Zhou, C. R., & Zhang, Z. Y., (2008). *Adv. Drug Deliv. Rev., 60*, 1650.
39. Sahu, S. K., Mallick, S. K., Santra, S., Maiti, T. K., Ghosh, S. K., & Pramanik, P., (2010). *In vitro* evaluation of folic acid modified carboxymethyl chitosan nanoparticles loaded with doxorubicin for targeted delivery. *J. Mater. Sci. Mater. Med., 21*(5), 1587–1597.
40. Guan, M., Zhou, Y., Zhu, Q. L., Liu, Y., Bei, Y. Y., Zhang, X. N., & Zhang, Q., (2012). N-trimethyl chitosan nanoparticle-encapsulated lactosyl-norcantharidin for liver cancer therapy with high targeting efficacy. *Nanomed., 8*(7), 1172–1181.
41. Kim, K., Kim, J. H., Park, H., Kim, Y. S., Park, K., Nam, H., et al., (2010). Tumor-homing multifunctional nanoparticles for cancer theragnosis: Simultaneous diagnosis, drug delivery, and therapeutic monitoring. *J. Control. Release, 146*(2), 219–227.
42. Mao, H. Q., Roy, K., Troung-Le, V. L., Janes, K. A., Lin, K. Y., Wang, Y., August, J. T., & Leong, K. W., (2001). Chitosan-DNA nanoparticles as gene carriers: Synthesis, characterization and transfection efficiency. *J. Control. Release, 70*(3), 399–421.
43. Choi, K. Y., Chung, H., Min, K. H., Yoon, H. Y., Kim, K., Park, J. H., Kwon, I. C., & Jeong, S. Y., (2010). Self-assembled hyaluronic acid nanoparticles for active tumor targeting. *Biomaterials, 31*(1), 106–114.
44. Choi, K. Y., Min, K. H., Na, J. H., Choi, K., Kim, K., Park, J. H., Kwon, I. C., & Jeong, S. Y., (2009). *J. Mater. Chem., 19*, 4102.
45. Yu, M. K., Park, J., & Jon, S., (2012). Targeting strategies for multifunctional nanoparticles in cancer imaging and therapy. *Theranostics, 2*(1), 3–44.
46. Platt, V. M., & Szoka, F. C., (2008). Anticancer therapeutics: Targeting macromolecules and nanocarriers to hyaluronan or CD44, a hyaluronan receptor. *Mol. Pharm., 5*(4), 474–486.
47. Lee, H., Lee, K., & Park, T. G., (2008). Hyaluronic acid-paclitaxel conjugate micelles: Synthesis, characterization, and antitumor activity. *Bioconjug. Chem., 19*(6), 1319–1325.
48. Min, H. S., Son, S., Lee, T. W., Koo, H. H. Y., Yoon, J. H., Na, Y., et al., (2013). Liver-specific and echogenic hyaluronic acid nanoparticles facilitating liver cancer discrimination. *Adv. Funct. Mater., 23*(44), 5518–5529.
49. Aynie, I., Vauthier, C., Chacun, H., Fattal, E., & Couvreur, P., (1999). Spongelike alginate nanoparticles as a new potential system for the delivery of antisense oligonucleotides. *Antisense Nucleic Acid Drug Dev., 9*(3), 301–312.

CHAPTER 5

Recent Advancement of Nanobiotechnology in Cancer Treatment

ABHINAV SHRIVASTAVA

Department of Biotechnology, College of Life Sciences, Cancer Hospital and Research Institute, Gwalior, Madhya Pradesh, India

**Corresponding author. E-mail: abhi.shri76@gmail.com*

ABSTRACT

Nanomedicine is an advance application of nanobiotechnology, which develops through a combination of medicine with nanotechnology. It is a novel and sophisticated technique against cancer through detection, prevention, therapeutic treatment with highly precision, and greater efficacy along with minimum side effects in comparison to standard medicine. Nanomedicine is a drug with very small sizes ranging from a few to many hundred nanometers. There are various ways of applications of anticancer nanomedicine like nanodrug delivery systems, nanoanalytical contrast reagents, and nanopharmaceuticals. Nanotechnology is being applied to cancer in two broad manners: the preparation of nanocarriers, like nanoparticles (NPs) that can be loaded with drugs and then targeted to particular tumors which show much better therapeutic effect against cancer due to their very fine size, target specificity, and modifiable nature and another method is development of nanosensor for detecting the cancer. This chapter discusses about nanomedicine-based clinical applications and their advantages in the advancements of cancer therapy.

5.1 INTRODUCTION

Nanobiotechnology is specially used for the prevention, therapeutic management, and healing of any disease through the combination of nano and construction of devices at the molecular level are known as nano-medicine. Nanooncology is the use of nanobiotechnology to the control of cancer management. It is a very significant application of nanomedicine. Meanwhile, nanomedicine is developing by the function of nanotechnology in the field of medicine [19]. New nanomedicines are preparing for treatment of cancer due to the following reasons:

1. Multifunctional approaches;
2. Improved effectiveness and multivalency;
3. Better targets selectivity;
4. Theranostic efficacy;
5. Pharmacokinetics becomes changes;
6. Production under control;
7. Kinetics and release of the drug if regulated properly;
8. New properties and communications;
9. Deficiency of immunogenicity;
10. Physical strength is improved [20].

Nanomedicines are generally used in the treatment of various cancers by drug delivery methods.

The major part of nanomedicine can be categorized into one of the following fields:

- Therapeutic drug delivery systems to deliver NPs at specific targets through specific routes;
- New biomaterials and tissue engineering (TE) for active tissue regeneration;
- Development of biosensors, biochips, and novel molecular materials for diagnostic and analysis purposes, monitoring, and imaging.

The basic principle of nanomedicine is enhancing the therapeutic index of anticancer medicine through altering the pharmacokinetics and tissue distribution so that the transportation of drugs toward the action site becomes improved [12]. There are many advantages of Nanotechnology-based drugs over conventional drugs, including half-life improvement, retention time, targeting capacity, various types of targeting ligands, and less or without side effects in patients (Table 5.1) [6].

TABLE 5.1 Differences Between Traditional and Nanomethods

Uses	Traditional Methods	Nano Methods
Detection or Diagnosis	Biopsy, FNAC, Immunehistochemistry FISH, etc., used for diagnostic purpose.	Raman probe and Quantum dots can be applied for diagnosis.
	Disease cannot detect at early stage.	Detection at early stage is possible.
	Detection by only in vitro is possible.	Detection by both types In vitro and in vivo is possible.
	Detection at real time is not possible.	Detection at real time is too possible.
Imaging	CT, MRI, X-ray, isotope, PET Scan can be applied.	Quantum dots can be applied.
	Ultrasonography is applied.	UCM ultrasonography can be applied.
		Very early diagnosis of neovascularization is also possible.
Early detection of cancer	Cytological study, FNAC, and biopsy are applied.	Vital optical imaging and also inexpensive mass screening is used.
Drug delivery	Drug delivery by oral, intravenous, intramuscular, intra-arterial drug delivery is used.	Vital optical imaging and also inexpensive mass screening is used.
		Targeted delivery can also be applied.
		Small dose delivery, low systemic toxicity, better, and rapid responses.

Many nanomedicines show an increased surface area and quantum effects in comparison to conventional drugs, so it can affect the size-dependent properties and also alter the physiological behavior of nanodrug. NPs are especially applied for the treatment of cancer and diseases of the central nervous system, because they are capable of passing the blood-brain barrier (BBB) [10]. Altering the size and structure of nanodrug or nanosystems, time of residence, and suitable drug release pattern can also achieve. Clinical purpose of anti-cancer nanomedicines can be broadly categorized into five main types: polymeric conjugates, liposomes, polymeric micelles, polymeric NPs, and others, although there is some overlap between categories [12].

Nanomedicine or nanosystems are of very small in size at the nanoscale and very high surface area in comparison to their volume. So these nanomedicines can easily interact with the cellular surface and intracellular part of the cell. Nanosystems have four distinctive features in comparison to other cancer medicine:

1. these nanosystems also have diagnostic or therapeutic characters and constructed in such manner that they can be designed to cart high drug load;
2. these nondrug or systems also joined to the ligand through multivalent manner, so attaché with target cells with high affinity and specificity;
3. nanomedicine or nanosystems can also prepare in such a way so they can also provide along with combinational therapy during cancer treatment; and
4. the use of nanodrug or nanosystems can avoid the risk of mechanisms of resistance generate due to the conventional drug [4].

The extravasation of any medicine or molecule into the tumor tissue and their holding time is called enhanced permeation and retention (EPR) effect. EPR develops a type of environment at the molecular level, which directing, attacking very specifically toward the lymph and blood vessels existing in tumors [14]. Nanomedicines have high EPR in compare to conventional normal medicine.

Many research groups are working on drug delivery systems. There are many types of Nanomedicines like dendrimers, caged NPs, etc., which may be applied to successful drug targeting in the right manner, according to type and severity of disease as well as a patient condition [24]. This system is also used for the treatment of fat cells, tumors, cancers cells, etc. [13].

There are two types of drug delivery systems are available:

1. Surface modification systems prepared to enhance cellular growth and to inhibit immunogenic reactions; and

2. Particle-based systems constructed to target the drug toward cells and tissues.

There are many examples of particle-based systems like gold, silver nanoparticles (AgNPs), peptides, etc. The nanocarriers with very small in size have the benefit that they easily enter and migrate in different sizes of blood vessels and uptake by cells. But the drawback is that the preparation of fine size NPs becomes difficult, and drug-loading capacity is also very poor [15].

Nano drug should have few following biological features [11]:

- it should be nonpoisonous and traceable;
- biodegradable in nature;
- they must be target specific for cell and tissue.

For different diagnostic and treatment in the nanoparticle must have multilayer coverings and also possess the following features [11]:

- very fine-sized of nanoparticle carrier;
- highly stable in media at physiological condition;
- attachment with a suitable drug;
- carrier system must have low or without toxicity;
- the drug should have rapid release and highly active;
- it should have "stealthiness" to omit the immune response.

There are various forms through which drugs may be released, like drug may be degraded, diffused, swelled, and interactions based on affinity such as gold particles. Generally, the drug is used ether in orally or through injection in the human body.

NPs are made up of polymers, which may be of two types: natural and artificially prepared or synthetic [4]. Natural polymers, like lipids, polysaccharides, proteins, etc., are non-toxic in nature, and they are biodegradable. But their use becomes limited due to variations in the degree of purity, and there is a need of proper optimization before encapsulation of drugs. Nowadays, synthesized polymers are used commonly, because their manufacturing is easy, and NPs constructed from it are biodegradable.

The major benefit of drug delivery methods for nanoparticle is that the discharge of the active component of the drug can be organized by the degradation of the nanoparticle shell constructed from a polymer. So this also leads to focuses on the target; due to the intracellular nature duration of delivery also extended, the drug dose is reduced, and resulting side effects also diminish, as well as medication and wellbeing of the patient is enhance [10].

NPs constructed for tumor therapy have various parts. Generally, one part is nanocarrier, and another is medicine or drug associated with it [13]. Drug-carrier containing NPs are assumed as colloidal systems that may behave as drug transporter, either in form of nanospheres (drug is distributed in a matrix system) or a type of nanocapsules (pools in which the drug is kept in hydrophobic or hydrophilic core enclosed by a single membrane made up of polymers) [13]. There are various types of Nanoparticle carriers generally consist of lipid-based carriers such as liposomes and micelles, iron oxides, biodegradable polymers, gold, dendrimers, organometallic compounds, etc. [16].

A significant advantage of some nanomedicines is the capability to prepare a drug without using dose-limiting toxic excipients according to formulations in the present market; tolerability is also improving and permitting more drugs to be directed to patients [3].

There are two types of drug delivery strategies:

a. active drug delivery manner; and
b. passive drug delivery manner [1].

Both types of strategies are used to increase the concentration of nanodrugs intracellularly in cancerous cells with minimal damage in normal cells, instantaneously anti-cancer activity is improving, and systemic toxicity is decreased [4].

1. **Drug Targeting Strategies:** There are two basic requirements to gain active drug delivery: (I) the drug should able to approach the desired location of tumor after administration with the nominal loss of its volume and action in blood circulation. (II) Drugs should be target-specific, i.e., it only kills tumor or cancer cells deprived of detrimental effects to healthy cells and tissue. These necessities may be assisted using two approaches: passive and active targeting of nanodrugs [19].

2. **Passive Targeting:** Passive directing of nanodrug and its accumulation in the vicinity of tumor tissue is a size-dependent procedure.

It causes a leaking vasculature and poor lymphatic drainage, which is a pathophysiological feature of the vessels of tumor. Due to leaking vasculature found in cancerous tissue make it possible that nanoparticle to migrate easily toward the cancerous tissue and kill them [3]. The extravasation of any substance into the tumor cells and tissue and their holding or retention is generally known as the EPR effect. So this process has the advantage of the improved permeability and retention (EPR) effect.

One problem with targeting of nanoparticle was encountered was that these particles were identified as extraneous or foreign bodies, maybe opsonized by the mononuclear phagocyte system (MPS), So the availability of the drug become lessen at the desirable site. Nowadays, well-prepared nanocarriers are available, which have a coating of polyethylene glycol, and this process is known as pegylation. These NPs have the capability to avoid arrest by the MPS. This drug delivery method is called as the stealth systems [21].

There are various methods include in passive targeting systems like gold particles, silica NPs, silver particles, liposomes, quantum dots (QDs), iron oxide nanoparticles (IONPs), hybrid NPs, and micelles [25].

Although passive targeting method is very effective, but there are some drawbacks, they suffer from several limitations like few tumors do not act according to the EPR effect, and some tumors are not well organized. So another method is an active drug targeting method, which is based on interactions with high-affinity molecules like antibodies, small molecule, aptamers, peptides, etc. that joined with specific cellular receptors [4].

3. **Active Drug Targeting:** Active targeting is an alternate method to overcome the restrictions of the passive targeting method. In this strategy, any particular ligand or antibody is associated with NPs. This method is considered very effective and specific for targeting any drug at the desired location [22]. Over-expression of receptors of the tumor cell membrane and mechanisms of phagocytosis/endocytosis also plays a very crucial role for site targeting and entry toward tumor cells and tissue to kill them [2].

4. **Choice of Target Receptor:** Construction of targeted NPs depends upon the design of crucially depends on, choice of suitable receptors or antigens in the vicinity present on cancer or tumor cells. Prominentlyidyllic receptors or antigens are those which are exclusively present only on cancer cells with high density but not on normal cells. Directed nanoconjugate must internalize subsequently interact with particular ligands to circumvent simple diffusion of the nanodrug proximate the cell surface. However, this drug transport process outside the cell may diffuse or redistribute the drug to the adjacent normal cells and tissues, in its place of the mass of cancer cells. Internalization of target receptor through receptor-mediated endocytosis causes the drug discharge pattern precisely toward cancer cells [22].

Careful and effective drug transport mechanisms to the cancerous cell and tissue are the emergent tasks. The selection of appropriate ligands has an important role in the mechanism of activating receptor-mediated endocytosis. Construction material for ligands can be natural constituents such as foliate and growth factors or any other similar biomolecules with lower molecular weight and lower in immunogenicity to antibodies. The latest improvements in molecular biology and genetic engineering permit improved antibodies or any other ligand to be used as directing moieties in an active-targeting method. Nowadays, monoclonal antibody or fragments of the antibody are the most commonly used ligands for such types of targeted remedies (Table 5.2) [6].

5.1.1 FDA APPROVED NANOMEDICINES: AN UPDATE IN CANCER THERAPY

Scientists have previously completed advancement through chemotherapeutic Nanomedicine in the hospital. Nanomedicines have revealed the potential for improving bioaccessibility, increasing solubility of medicine, active drug targeting, and loading of high drug concentration [5].

Numerous compounds that are in different phases of trials or at present permitted by the US Food and Drug Administration (FDA) are stated in Table 5.3 [3].

TABLE 5.2 Briefing of Nanoparticles-Based Nanomedicine Used for Treatment of Cancer [5, 10, 23]

Types	Characteristics	Uses
Polymer	Few are biodegradable	Controlled manner (triggered)
Dendrimer	Biocompatible, cargo, Low polydispersity	Gene therapy, drug delivery
Lipid	Biocompatible can carry hydrophobic cargo, typically 50–500 nm in size	Drug delivery
Quantum dots	Broad excitation no dots photobleaching, tunable emission, typically 5–100 nm	Optical imaging
Gold	Biocompatibility, typically 5–100 nm	Hyperthermia therapy, drug delivery
Silica	Biocompatibility	Drug delivery (encapsulation), contrast agents
Magnetic	Super paramagnetic, ferromagnetic (small reminisce to minimize aggregation), super ferromagnetic (~10 nm), paramagnetic	Contrast agents (MRI), hyperthermia therapy
Nanobot	Nanoelectromechanical systems, biocompatible	Drug delivery, diagnostics, imaging
Carbon-based	Biocompatible	Drug delivery, sensor
Single crystal nanowires	Biocompatible	Diagnostics and implanted sensors, drug delivery, medical devices.

5.2 ADVANCED TECHNOLOGIES BASED ON NANOTECHNOLOGY

In the cancer treatment, nanodrug treatment is necessary due to specificity toward cancer cells but not damage the normal cells. These therapies are very latest and non-invasive in nature. Bioengineered NPs or nanosystems are novel weapons in cancer therapy, which comprises radiotherapy, and radiofrequency therapy, photodynamic therapy (PDT), gene therapy, and theranostics [6]. After the application of nanotechnological methods, tumor cells are killed without the destruction of normal cells.

TABLE 5.3 Various Anti-Cancer Nanomedicines With Pharmaceutical Description

Type of Nanomedicine	Medicine or Drug	Name of Product /Brand Name/Company Name	Clinical Manifestation	Type of Phase
Polymeric nanoparticles	AZD2811 (AZD1152 hydroxyquinazolinepyrazol anilide; Aurora-B Kinase Inhibitor)	AZD2811 (Accurin™)	Nanoparticle/AstraZeneca advanced solid tumors	Phase I
	Docetaxel + Prostate-specific membrane antigen (PSMA)	BIND-014 (Accurin™)/BIND therapeutics	Cholangiocarcinoma, cervical cancer, bladder cancer, head and neck cancer, non-small cell lung cancer subtypes	Phase II
Liposomes		Doxil™/Janssen	Kaposi's sarcoma, ovarian cancer, multiple myeloma	Approved
		ThermoDox™/Celsion	Primary hepatocellular carcinoma, refractory chest wall breast cancer, colorectal liver metastases	Phase III Phase II
	Doxorubicin	Teva UK / Myocet™	Metastatic breast cancer	Approved
		2B3–101/2-BBB medicines BV	Brain metastases Glioma	Phase II
	Vincristine	Marqibo™/spectrum Pharmaceuticals	Acute lymphoblastic leukemia	Approved
	Irinotecan	Onivyde™/Merrimack Pharmaceuticals	Metastatic pancreatic cancer (2nd line)	Approved
			Gastric cancer	Phase II
	Cytarabine	Depocyt™/Pacira pharmaceuticals	Lymphomatous meningitis	Approved
	DACH-platin	NC-6004 Nanoplatin™/ Nanocarrier™	Pancreatic cancer, head and neck cancer, non-small cell lung cancer, bladder cancer	Phase III

TABLE 5.3 *(Continued)*

Type of Nanomedicine	Medicine or Drug	Name of Product /Brand Name/Company Name	Clinical Manifestation	Type of Phase
	Cisplatin	Regulon /Lipoplatin	Non-small cell lung cancer	Phase III
	Daunorubicin	Galen /DaunoXome™	HIV-related Kaposi's sarcoma	Approved
	Ratio of Cytarabine: Daunorubicin 5:1	CPX-351/Celator	Acute myeloid leukemia	Phase III
Polymeric conjugates	Asparaginase	Oncaspar™ (PEG)/Baxalta	Acute lymphoblastic leukemia	Approved
	Irinotecan	NKTR102 (PEG)/Nektar	Metastatic breast cancer	Phase III
	Camptothecin	CRLX101 (nanoparticle)/ Cerulean	Renal cell carcinoma (3rd/4th line), ovarian cancer (2nd/3rd line)	Phase II
	Paclitaxel	CTI Biopharma/Opaxio™	Ovarian cancer, non-small cell lung cancer (female)	Maintenance of Phase III
				Phase II
	Camptothecin	CRLX101/Cerulean	Renal cancer, small cell lung cancer, ovarian cancer	Phase II
	Diaminocyclohexane (DACH) platinum	AP 5346 ProLindac™ / (Hydroxypropylmethacrylate)	Cancer of ovary	Phase II
Other	Paclitaxel	Abraxane ™/Celgene	Advanced breast cancer, advanced non-small cell lung cancer, advanced pancreatic cancer	Approved
	Tumor necrosis factor (TNF)	CYT-6091/CytImmune	Non-small cell lung cancer	Phase II
	Irinotecan HA-irinotecan	HyACT™/Alchemia	Colorectal cancer	Phase II
			Lung cancer	Phase III

5.3 CONCLUSION

Nanobiotechnology is the application of nanotechnology in medicine, which rapidly growing the field of interdisciplinary research and involves in diagnosis, treatment, and/or screen any disease at the nanoscale. Cancer nanotechnology is advance biomedical uses of NPs, which are nano level and that are capable of overwhelmed biological obstacles, especially, identify a particular kind of cancer cell, and collect favorably in tumors. NPs have various therapeutic applications like; they have the capability to deal with innovative methods of noninvasive tumor detection, diagnosis, analysis, and treatment. There are many varieties of ligands directed against cancer cells, such as peptides, antibodies, or small molecules, which can be joined with NPs for targeting of tumor-specific antigens and vasculatures substances with great specificity and high affinity. Moreover, diagnostic agents and chemotherapeutic medicines can be combined into their design for more proficient imaging and therapeutic effect of the tumor with less or no side effects. The latest improvements in nanomedicine increase exciting opportunities for forthcoming applications of NPs in personalized health care management of cancer therapy. Nano- technology has become an empowering expertise for personalized oncology, in which tumor molecular profile of every person and genetic and/or molecular markers for predic- tive oncology are applied to the prediction of disease development, advancement, and therapeutic outcomes [6].

Nanomedicines can be delivered in two ways: passive and active manner. Appro- priately constructed NPs have the ability to gather in tumors either by passive or active targeting methods and increase the cytotoxic properties of antitumor drugs. Numerous anticancer nanodrugs formulations have been assessed, and only a few medicines are at present accepted for clinical purposes, and other nanodrugs are undergoing various phases of clinical trials [17].

There are many types of nanocarriers like dendrimers, liposomes, micelles, PEG, gold; magnetic NPs, silica NPs, and QDs have many benefits [8]. Nowadays, bioengineered NPs based on new techniques like radiotherapy and radiofrequency therapy, PDT, gene therapy, and theranostics are developing, but further sophistica- tion is necessary.

FDA has developed and approved various nanodrugs against many types of cancers, although some drugs are in different phases of clinical trials. Upcoming opportunities for nanomedicines are looking towards delivering the subsequent generation of nanodrugs: molecularly targeted agents, substances which induces apoptosis, treatment therapy based on DNA/RNA, combinations of various drugs, peptides, etc. There are many barriers in the targeting of nanodrugs include the considerable gathering of the drug around the target, the passage of nanodrug through the cell membrane. To gain the synergistic effect of many drugs ratios toward the target, a low therapeutic index and high cost of the drug also. So these hurdles should be overcome by the use of unique technology of nanomedicine and knowledge.

KEYWORDS

- **anticancer**
- **drug**
- **nanocarriers**
- **nanomedicine**
- **nanoparticles**
- **photodynamic therapy**

REFERENCES

1. Verma A., (2015). Article on latest trends in nanomedicine. *Journal of Nanomedicine Research, 2*(2), 2015.
2. Bamrungsap, Z. Z., Tao, C., Lin, W., Chunmei, L., et al., (2016). *Therapeutics, 7*(8), 1253–1271.
3. Danhier, F., Feron, O., & Preat, V., (2010). To exploit the tumor microenvironment, passive and active targeting of nanocarriers for anticancer drug delivery. *J. Control Release, 148*, 135–146.
4. David, E. R., (2009). *Bionanotechnology: Global Prospects*. CRC Press, Taylor & Francis Group, Boca Raton.
5. Dawidczyk, C. M., Kim, C., Park, J. H., Russell, L. M., Lee, K. H., et al., (2014). State-of-the-art in design rules for drug delivery platforms: Lessons learned from FDA-approved nanomedicines. *J. Controlled Release, 187*, 133–144.
6. Deshpande, G. A., (2016). *Cancer Nanotechnology: The Recent Developments in the Cancer Therapy, 1*(1), pp. 1–6, 555551. https://pdfs.semanticscholar.org/933d/832ae fc044acf6ce6a8134b302a00d5e0c7f.pdf
7. Dey, N. S., Majumdar, S., & Rao, M. E. B., (2008). Multiparticulate drug delivery systems for controlled release. *Trop. J. Pharm. Res., 7*(3), 1067–1075.

8. Drbohlavova, J., Chomoucka, J., Adam, V., Ryvolova, M., Eckschlager, T., Hubalek, J., & Kizek, R., (2013). *Current Drug Metabolism, 14*, 547–564.
9. Ferrari, M., (2005). Cancer nanotechnology: Opportunities and challenges. *Nat. Rev. Cancer, 5*(3), 161–171.
10. Tossi, G., Costantino, L, Ruozi, B., Forni, F., & Vandelli, M.A., (2008). Polymeric nanoparticles for drug delivery to central nervous system. *Exp. Opin. Drug Deliv., 5*(2), 155–174.
11. Gregory, F. S., Vladimir, S., Karel, U., & Gero, D., (2009). Multifunctional cytotoxic stealth nanoparticles. A model approach with potential for cancer therapy. American Chemical Society. *Nano Letters, 9*(2), 636–642.
12. Hare, J. I., et al., (2016). Challenges and strategies in anti-cancer nanomedicine development: An industry perspective. *Adv. Drug Deliv. Rev.*, http://dx.doi.org/10.1016/j.addr.2016.04.025 (Accessed on 5 November 2019).
13. Juillerat-Jeanneret, L., (2008). The targeted delivery of cancer drugs across the blood-brain barrier: Chemical modifications of drugs or drugnanoparticles? *Drug Discov. Today, 13*(23&24), 1099–1106.
14. Kumar, R., Roy, I., Ohulchanskky, T. Y., Vathy, L. A., Bergey, E. J., Sajjad, M., & Prasad, P. N., (2010). *ACS Nano, 4*, 699–708.
15. Maja, P., & Rudolf, P., (2010). Nanomedicine: A way of targeting and detection of cancer cells. *Seminar Proceedings*, pp. 1–14.
16. Mishra, B., Patel, B. B., & Tiwari, S., (2010). Colloidal nanocarriers: A review on formulation technology, types and applications toward targeted drug delivery. *Nanomed.-Nanotechnol. Biol. Med., 6*(1), 9–24.
17. Pillai G., (2014). Nanomedicines for cancer therapy: An update of FDA approved and those under various stages of page 4 of 13 development. *SOJ Pharm. Pharm. Sci., 1*(2), 13.
18. Rajshri, M. N., & Tarala, D. N., (2007). Application of nanotechnology in biomedicine. *Indian Journal of Experimental Biology, 45*, 160–165.
19. Robert, A., & Freitas, Jr., (2005). *What is Nanomedicine, Nanomedicine: Nanotechnology, Biology, and Medicine, 1*, 2–9. Elsevier Inc.
20. Scheinber, D. A., Villa, C. H., Escorcia, F. E., & McDevitt, M. R., (2010). Conscripts of the infinite armada: Systemic cancer therapy using nanomaterials. *Nature Reviews, 7*, 266–276.
21. Torchilin, V. P., (2007). Targeted pharmaceutical nanocarriers for cancer therapy and imaging. *AAPS J., 9*(2), E128–E147.
22. Trochilin, V. P., (2010). Passive and active drug targeting: Drug delivery to tumors as an example. *Handbook Experimental Pharmacology, 197*, 3–53.
23. Ahmad, U. & Md. Faiyazuddin, (2016). Smart nanobots: The future in nanomedicine and biotherapeutics. *J. Nanomedicine Biotherapeutic Discov., 6*, 1. https://www.longdom.org/open-access/smart-nanobots-the-future-in-nanomedicine-and-biotherapeutics-2155-983X-1000e140.pdf
24. Wang, N. X., & Von Recum, H. A., (2011). Affinity-based drug delivery. *Macromol. Biosci., 11*(3), 321–332.
25. Wang, X., Yang, L., Chen, Z. G., & Shin, D. M., (2008). Application of nanotechnology in cancer therapy and imaging. *CA Cancer J. Clin., l58*, 97–110.

CHAPTER 6

Role of Nanoparticles in Common Cereals

RAGHVENDRA KUMAR MISHRA, ARUNCHAND RAYAROTH, and RAJESH SINGH TOMAR

Amity Institute of Biotechnology, Amity University Madhya Pradesh, Gwalior, India

**Corresponding author. E-mail: rkmishra@gwa.amity.edu (R. K. Mishra); arunrayaroth@gmail.com (A. Rayaroth); rstomar@amity.edu (R. S. Tomar)*

ABSTRACT

Nanotechnology is considered as one of the upcoming technologies with potential to revolutionize multiple fields of application including one of the most demanding future field, i.e., agriculture. Numerous studies have tried to explore the potential of nanotechnology in improving the crop yield especially of the major cereals including maize, rice, and wheat. Most of the studies have shown positive outcomes with nanoparticle application including improvement in growth, nutritional value, crop yield, controlling diseases, resistance to diseases and adverse environmental conditions, as a biofertilizer, etc., among the reported. Specialized process aids such as tribo-mechanical activation to produce specialized nanoparticle-based products like Herbagreen with high efficiency to improve crop quality especially of maize is quite promising. Further, the ability of crops like maize to grow especially under water stressed conditions in presence of nanoparticles and the potential of Nano-TiO_2 to cleanse wastewater of heavy metals are propitious. Research has also reported the adverse effects of nanoparticle use including cytotoxicity growth stunt, etc., which was majorly because of the excess use of nanoparticles. However, the reports highlighting the economic benefits of nanoparticle application to improve yield is very less. Further studies are required to prove it commercial viability as well as its effect in substantially improving the crop yield to an extent where its application can be practical is the need of hour.

6.1 INTRODUCTION

Nanotechnology is considered as one of the emerging technology of the century. With applications ranging from medicine, electronics, agriculture, industrial, and even interestingly into art, nanotechnology opens up a wide arena of opportunities to enhance feasibility, yield, process efficiency, profitability, and quality. Discovery of various innovative nanochemical moieties with novel physical and chemical properties have found multidisciplinary application due to its uniqueness. Rapid growth in various sectors is demanding the development of innovative ideas from this segment of science, which already has taken a giant leap towards realizing various concepts. Agriculture is one such challenging sector, which requires a major drive due to ever-rising demand for food and feed in a world, which already is under the threat of famine and massive drought. The past few decades have seen various technological innovations in this sector, including usages of mobile applications for farm management, hydroponics: the science of water management, drones for farm monitoring, and also the development of GM crops, which is still under lot of controversies.

Nanotechnology, the engineering of functional systems at the molecular scale, is now gaining popularity in the agricultural sector due to several successful and positive responses from different agricultural subsectors. Even though studies have found productive results for the application of nanotechnology in agriculture, its real-life application still needs to be extensified. Making the technology more economical, popularizing and creating awareness among the common man, while retaining the theoretical definition in the delivered products along with the guaranteed claim, is some of the major challenges to be addressed under the current scenario. Some of the major agricultural applications include use of nanoparticles (NPs) in controlling the plant diseases, nanosensors, and diagnostic devices including carbon nanotubes (CNTs), mesoporous silica nanoparticles (MSNs) for water purifying, plant protection devices including nanoemulsion, silver nanoparticles (AgNPs), and bio-beads for mediating gene transfer.

6.2 ROLE OF NANOPARTICLES (NPs) IN MAIZE

Maize, being a major protein and calorie source, is considered an important food source, especially for the large community in the developing world in its different processed forms. Maize provides about 30% of daily protein intake and 40% of daily recommended energy intake. Apart from being a major protein source for humans, maize, along with soya, forms a major and widely used ingredient in feed and fodder across the world. The nutritive quality, mainly protein, of this global grain has been improved over the years through various genetic-based breeding as well as different processing strategies. Nanotechnology has nevertheless been applied at mainstream or application-level on this crop for improvement purposes.

Multiple researches have been studied carried out on the interaction of NPs with different plant and their effect on growth and development. Metal NPs are some of the most common NPs extensively researched, with results enlightening its positive and negative effects based on usage levels. Titanium dioxide ($nTiO_2$), silicon dioxide ($nSiO_2$), cerium dioxide ($nCeO_2$), magnetite (nFe), aluminum oxide (nAl_2O), zinc oxide (nZnO) and copper oxide (nCuO) are some of the most common metal oxide NPs explored to understand its influence on plant growth and development and even phytotoxicity. Germination of maize seeds are generally not found to get affected by the commonly used metal oxide NPs, whereas the root elongations are found to be significantly inhibited mainly by nCuO and nZnO. The phytotoxic effects on roots are generally found to be concentration-dependent and also size-dependent with larger molecules showing higher toxicity compared to their smaller counterparts. Even though phenotypic studies on the underlying toxicity mechanism involving NPs have been widely conducted, studies targeting the physiological, metabolic, and genetic aspects are still in its infancy [1].

Nanotechnology is undoubtedly one of the breakthrough inventions of the modern era with applications ranging from agriculture to other prime areas of science, including biotechnology, medicine, etc. The most crucial part of nanoscience is the technology explored to bring out nanoparticles, which can greatly make the difference. Tribo-mechanical activation is one such process through which a specialized Herbagreen nanoparticle was created, which has the capability to directly penetrate plants. The technology has been successfully applied in maize farming with the highest protein level observed in the Herbagreen group [2].

Nano biofertilizer is another novel application which in combination with mineral NPs has found application, especially in stress conditions. Maize has been cultured successfully under water stress condition on loam clay soils with significant results, including the maximum chlorophyll SPAD value, 100-grain weight as well as high harvest index. In a similar nano-based study, increase in chlorophyll content of maize leaf was found on the application of iron sulfate in soil and foliar spray (2.1. Amanulla). Such a combination application can also be used to reduce the irrigation period as well as a positive influence on growth and yield components [3].

Nano zinc and iron are one of the most common and widely used mineral NPs with successful trials being conducted for seed priming and foliar application, especially on forage corn both at field level and greenhouse conditions. Trials have shown that these nanoparticles have to lead substantial development in plant growth and development, including plant height, total dry biomass, crude protein, soluble carbohydrates, and phosphorus uptake along with significant improvement in leaf chlorophyll concentration. Zinc NPs application has also found increased emergence percentage suggesting the better ability of seedlings to grow and withstand adverse environmental conditions, whereas iron was demonstrated to have a great role in the synthesis of chlorophyll, photosynthesis improvement, and plant growth regulation. This may be due to the involvement of iron in the enzyme coproporphyrinogen oxidase, involved in the biosynthesis of d-aminolevulinic acid and its role in the synthesis of chloroplastic mRNA and rRNA, which control chlorophyll synthesis. With significant roles of Zinc and Iron in cell elongation, synthesis of tryptophan, chlorophyll, and many other indirect involvements in growth and development factors, these NPs can play a pivotal role in the future of culturing maize and other demanding crops. NPs are also found to increase the nutrient uptake, and protein and carbohydrate concentration compared to their parent forms. This may be due to the improvement of solubility and dispersion of insoluble nutrients in the soil, reduction in soil absorption and fixation, and enhancement of bioavailability by the nanosized formulation of mineral micronutrients. This is again attributed to the high surface area to volume ratio, high solubility, and specific targeting due to small size, high mobility, and low toxicity of NPs.

NPs are typically found to be efficiently transported via carrier proteins through aquaporin, ion channels, and also found to be transported into the plant by forming complexes with membrane transporters or root exudates.

Recent findings also highlight the possibility of entry through stomata or the base of trichomes in leaves, the exact mechanism of nanoparticle uptake by plants is yet to be elucidated [4].

Indirect application of Nano-TiO_2 in maize cultivation through exploring the potential of this nanoparticle to cleanse the wastewater of heavy metals has been quite successful. Significant increase in shoot fresh and dry weight, root fresh and dry weight, as well as root area, chlorophyll a, chlorophyll b, and carotenoids content, were found on using nano-treated wastewater in culturing maize. The same study also found that the higher concentration of nano-TiO was phytotoxic to maize. Thus wastewater can be used and advised for use as an irrigation source for culturing maize since lower concentrations of nano-TiO did not have any toxic effects on its growth [5]. In another study, the effects of AgNPs was observed on plant growth and development. The various growth parameters, such as length of shoot and root, the surface area of the leaf, the concentration of primary metabolites were significantly probed in common bean and corn [6].

Magnesium oxide nanoparticles (MgO NPs) synthesized by biocombustion of aqueous leaf extracts are also found to be effective in maize plants. On the treatment of maize seeds with synthesized MgO NPs, the same had shown better germination, especially when the seeds were treated with synthesized MgO NPs. Studies focusing on physical growth characteristics of the maize plants like germination percentage, germination index, vigor seedling index, relative seed germination, relative root growth, etc., were also found to substantially improve the overall plant growth. Further, this nanoparticle can also substitute harmful fertilizers and pesticides by selectively inhibiting harmful fungi and bacteria present on the seed. A probable reason for many of the beneficial actions of MgO NPs on plants may be due to its ability to increase plant hormonal activities [7].

Statistical analysis on the influence of zinc nano-chelate foliar and soil application on morpho-physiological characteristics of maize (*Zea mays* L.) had shown a profound influence on multiple growth and development parameters. The nano-chelate zinc, when supplemented through the soil, showed a significant effect on plant height, 100-grain weight, and harvest index, whereas foliar application had almost lead to cent percentage more seed yield per plant compared to control. Further, the harvest index was observed to increase markedly when nanoparticle application was double directed towards soil as well as foliar level, which was not observed in case of grain weight and seed yield [8].

Titanium dioxide is yet another key nanoparticle found to have influenced the growth characteristics of *Zea mays* L. Factorial experiments were conducted with two treatment factors applied through spraying, one at the stage of plant growth. Significant improvements were found on chlorophyll content (a and b), total chlorophyll (a + b), chlorophyll a/b, carotenoids, and anthocyanins by using nano TiO_2. The maximum amount of pigment was recorded from the treatment of nano TiO_2 spray at the reproductive stage in comparison with control. The data also indicated that nano TiO_2 could noticeably facilitate an increase in crop yield, especially corn yield. Additionally, it was very evident from the data that nano TiO can positively influence the production and enhancement of *Zea mays* pigments, thereby increasing its market value, especially in the feed market. The higher amounts of pigments were obtained, especially by spraying of nano TiO_2 in the reproductive stages of the plant in comparison with spraying TiO (bulk) and distilled water (control) spraying in the vegetative stage. This may be because of, remobilization of photosynthetic material from leaves to fruits at flowering time. Nano-TiO also could have affected on photoreduction activities of photosystem II and electron transport chain and thereby increasing the levels of pigments. These changes further lead to increased photosynthetic efficiency; stimulated rubisco and also increased photosynthesis. In summary, it was recorded that the use of TiO_2 NPs would increase the yield of Zea mays [9].

Zinc oxide nanoparticles have also been compared with other physically different forms of Zinc oxide to understand their influence on maize. Three different physical forms of ZnO, including ZnO nano-colloid, micrometric ZnO other than their nano-counterpart, were introduced into irrigation water and were then used to supplement mineral poor soil. Their effect as studied on the growth of single cross 704 (SC704) corn at 2 ppm, was reported to be quite productive. The results showed that the addition of all three ZnO particle types in irrigation water improved shoot dry matter and leaf area index. The best results came from the ZnO nanoparticle treatment, which increased the shoot dry matter and leaf area indexes by 63.8% and 69.7%, respectively. Based on the experimental results, it was concluded that zinc NPs could improve corn growth and yield in mineral poor soils [10].

6.3 ROLE OF NANOPARTICLES (NPS) IN RICE

Rice is one of the staple foods in many parts of the world, especially in humid and sub-humid Asia, with India among the leading producer and consumer. Demand for rice and rice products have seen ever-rising growth in the past few decades, a drastic increase in population being a major reason. Technological adoption to improve the production of rice has been growing with IR8 has the first high yielding variety released by the International Rice Research Institute.

Many studies related to the usage of nanotechnology to improve production and yield of rice have been reported. Cerium, zinc, iron, silver, etc., are some of the most common minerals, the nano counterpart of which has been studied for their efficiency to positively influence on rice crop yield.

One such study involved used cerium oxide NPs (nCeO) to understand the physiological impacts of nano-minerals on rice seeds and plants. The research, which aimed to study the impact of nCeO on the oxidative stress and antioxidant defense system in germinating rice seeds, used rice seeds to germinate in $nCeO_2$ suspension at varying concentrations. As generally observed in mineral and nano-mineral supplementation, concentrated supplementation leads toxicity manifestation, including enhanced membrane damage and photosynthetic stress in the shoots. Other toxicity indications included enhanced electrolyte leakage and lipid peroxidation in the shoots of seedlings grown at 500 mg supplementation. The phytotoxicity was preceded by biochemical changes, including altered enzyme activities as well as levels of ascorbate and free thiols. At lower concentrations, even though the Cerium concentrations increased substantially proportionally to increasing supplementation, any visible signs of toxicity were not demonstrated in the seedlings [11].

Rice germination ratio, root length, shoot length, and fresh weight are some of the most common parameters investigated to understand phytotoxicity in plants. In such a targeted study, nanomolecules including Fe_2O short nanorods, Fe_2O_3 long nanorods, TiO_2 nanotubes, FeNPs; and MWCNTs was selected as the nano-moieties which can have a probable influence on rice (*Oryza sativa* L.). The research outcome showed that all the NPs under analysis inhibited the germination of rice seeds. The study, which showed a comprehensive understanding of NPs toxicity during plant germination, demonstrated significant promotion of root length with

Fe nanocubes, Fe_2O_3 short nanorods, Fe_2O_3 long nanorods, and TiO NPs and stimulated shoots growth at most concentrations, but without any obvious effect on fresh weight [12].

AgNPs with the advent of the latest advances in nanotechnology has been applied in many industries, including agriculture. The possibility of its usage did reach new heights, especially one with many positive indications of its usage in many agricultural crops like rice. Even on its wider usage for various industrial applications, the safety factor involved on uncontrolled usage still remains poorly evaluated. Various studies have found that increased molecular size as well as rising nanoparticle concentrations had a negative impact on crops especially rice. These results thus strengthen the understanding of environmental safety information with respect to nanomaterials [13].

Phytotoxicity was also found during the usage of one of the most common and widely used nano-molecule, i.e., nano-ZnO. A study conducted on rice (*Oryza sativa* L.) roots was aimed to understand the influence on growth characteristics, including seed germination percentage, root length, and the number of roots. The presented data did reveal the detrimental effects on rice roots at the early seedling stage but without any reduction in the percentage seed germination. Further, nano-ZnO was found to stunt roots length and reduce the number of roots. The study shows significant phytotoxic effects on rice plants upon direct exposure to specific types of NPs, and also emphasizes on the need for ecologically responsible disposal of wastes containing NPs. The study also highlights the necessity for further study on the impacts of NPs on agricultural and environmental systems [14].

Yet another research planned to evaluate nanoecotoxicity did propose a new plant model starting from the rice plant. The model was drafted to consider the impact of engineered NPs (Ag, Co, Ni, CeO_2, Fe_3O_4, TiO) on rice plants that were weakened by infections of *Xanthomonas oryzae* pv. oryzae bacteria. On completion of the study, it was found that some NPs increased the price sensitivity to the pathogen while others decrease the virulence of the pathogen. In addition, no enrichment in component metal concentration was detected in organs of rice, with the exception of Ni-NPs treatment [15].

Rice blast disease, caused by *Magna-porthe grisea*, is the most serious biotic threat to rice (*Oryza sativa* L.) production worldwide. It causes severe yield losses in Egypt, especially in epidemic years. The fungus is highly variable, so disease control is a challenge. The effect of AgNPs (20–30 nm) against rice leaf blast fungus evaluated under different cultivation conditions both in vitro and in vivo to the culture of *M. grisea* showed significant inhibition of both hyphal growth and number of colonies formed in a dose-dependent manner. The number of spores/ml decreased within all treatment. Under greenhouse conditions, AgNPs sprayed in varying concentrations of rice seedling leaves at three different intervals lead to leaf damaged. The damaged leaf area percentage (DLA%) indicated that the application of 100 ppm AgNPs was highly efficient before and after inoculation compared to the untreated plants of 80%. The detrimental effect of silver nanoparticle on fungal mycelial growth was further evident from the scanning electron microscopy results [16, 17].

6.4 ROLE OF NANOPARTICLES (NPS) IN WHEAT

Wheat is the most grown food crop and the second most-produced cereal with an overall world production of over 700 million tonnes. Even though wheat, as aforementioned, is considered as one of the largest cultivated crops, the gap between production and utilization is quite narrow. Considered as one of the most nutrient-rich cereal with amble concentrations of proteins, vitamins, minerals, fiber, etc., the conventional techniques used for wheat production are not keeping pace with the ever-rising demand. Wheat is another major crop on which nanotechnology has been extensively applied to understand its influence in enhancing the yield and nutritional value [18].

Silver, copper, and iron NPs are some of the most common mineral nanoforms used to experiment on common cereal groups to understand its influence on plant growth and development. The other NPs, like TiO_2-NP may get accumulated in plants and partially affect plant growth and development [19]. The effect of the exposure of wheat (*Triticum aestivum*) seeds to silver, copper, and iron NPs on germination and seedling vigor index has been studied under laboratory conditions. Seeds, when exposed to the NPs under various conditions, showed a reduction in germination percentage on exposure to silver and copper NPs, whereas germination recorded the maximum percentage on the application of iron NPs. Similarly, the root and shoot growth, which enhanced under iron nanoparticle application, showed a severe reduction on exposure to copper NPs. So copper has inhibitory while iron has a stimulatory effect on wheat germination and growth [20].

Titanium dioxide NPs have been shown to improve the growth parameters of multiple cereal grains at laboratory and farm levels. Nano-TiO experimented on wheat (*Triticum aestivum* L.) plants at seven varying concentrations of nano-TiO showed that germination rate and weighted germination index among the tested wheat germination indices were affected by TiO nanoparticle treatments. The lowest and the highest germination rate were obtained in control and 1000 and 1200 ppm concentration of nano-TiO treatments, respectively. In addition, plumule and radicle length, seedling fresh weight, and seedling vigor index showed a significant difference by nano-TiO concentrations. Plumule and radicle lengths at 1200 ppm concentration of nano-TiO were higher than the untreated control. The study showed that using TiO NPs in suitable concentrations caused increased seed germination of wheat cv. Parsi in comparison to control plants, with lower concentrations having an inhibitory effect on wheat germination characteristics [21].

Metal NPs can have a predominant influence on the physiological properties of plants, including seed germination, growth, and metabolism. In a study conducted to recognize the toxic effects of AgNPs and silver ions, callus cells of two varieties of wheat (*Triticum aestivum* L.): stress-tolerant—Parabola; stress-sensitive—Raweta were selected. Stress-induced by silver particles or ions (0, 20, 40, 60 ppm) were then investigated using different parameters such as morphological characteristics, lipid peroxidation, and mobilization of the defense system. The parameters such as the activity of antioxidant enzymes, glutathione (GSH), and proline contents were also measured. Further microscopic observations revealed the deformation of cells after treatment by the solution of higher silver concentrations. Even though malondialdehyde content in both studied varieties increased substantially, upsurge in proline content was observed mainly in the silver-treated cells. There was no effect of silver on the superoxide dismutases activity, while the activity of catalase was significantly decreased. The changes in the activity of peroxidases in both varieties were the opposite. The highest content of intracellular GSH was noticed at a concentration of 20 ppm of both AgNPs and silver ions. The results as reported demonstrated a significant similarity between the effects caused by the studied stressors: AgNPs and silver ions. The results characterized the mechanism of action of nanosilver on wheat callus: morphology disorder, damage to cell membranes, severe oxidative stress, and in consequence intensification of production of non-enzymatic antioxidants [22].

AgNPs were also experimented on hydroponic plant growth studies indicating probable phytotoxicity. Experiments were conducted on commercial Ag NPs (10 nm) to evaluate its phytotoxicity on a sand growth matrix. Both NPs and soluble Ag were recovered from water extracts of the sand after the growth of plants challenged with the commercial product. The Ag NPs reduced the length of shoots and roots of wheat in a dose-dependent manner with further impact on as shown by the branching in the roots of wheat (*Triticum aestivum* L.), thereby affecting plant biomass. The findings demonstrated the potential effects of environmental contamination by Ag NPs on the metabolism and growth of food crops in a solid matrix [23].

In an attempt to understand the environmental impact of engineered NPs on the environment, physiological and biochemical changes on the exposure of wheat plants to ferric nanoparticles where studied. The study, which was conducted under hydroponic conditions, evaluated the physiological effects and possible cell internalization of various concentrations of citric acid-coated-Fe_3O_4 NPs on wheat (*Triticum aestivum* L.) plants. Visualization of root sections by transmission electron microscopy (TEM) showed that Fe_3O NPs entered the root through the apoplastic route and were then detected in the root epidermal cell walls. Moreover, Fe_3O_4 NPs did not affect the germination rate, the chlorophyll content, and the plant growth, and they did not produce lipid peroxidation, nor alter O or H_2O accumulation with respect to control plants. The preliminary results showed that these Fe_2O_3 NPs are not phytotoxic, suggesting that they could potentially be useful for designing new products for agricultural use [24].

Advancement in nanotechnology and its impacts have raised concerns about the application of engineered nanomaterials (ENMs) in agriculture and the environment, especially on plants. In recent work, the effects of TiO_2 NPs (TiO_2 NPs) on the growth of wheat plants (*Triticum aestivum*) were investigated. The study, which was aimed with the main focus to evaluate the effects of TiO_2 NPs on plant's morphological parameters like root, shoot length, and biomass were conducted by sowing wheat seeds on soil containing varying concentrations of the nanoparticle. The experiment, which was statistically planned, used SEM-EDX analysis to observe the uptake of TiO_2 NPs by the wheat plants. The results showed that root, shoot length, and biomass were significantly affected by TiO_2 NPs treatments. An increase in the plant's root and shoot lengths and biomass was observed up to 60 mg/kg of TiO_2 NPs. Further increase of the nanoparticle concentration, negatively affected the root and shoot lengths and also a reduction in biomass [25].

Magnetite NPs are comparatively uncommon nano molecules experimented on agriculture crops compared to Zinc, Copper, iron, or their silver counterparts. Citrate-coated magnetite NPs have been experimented to assess its uptake by wheat plants as well as to understand its effect on the bioaccumulation and toxicity of individual and joint Cd^{2+} and Cr^{6+} levels. In the study, quartz-sand was used as the plant growth substrate. The influence was approximated by measuring the growth characters such as seed germination, root, and shoot lengths, and heavy metal accumulation. Magnetite, which exhibited very low toxicity, regardless of its uptake and distribution into roots and shoots, also increased plant root length. In contrast, the root length of wheat seedlings exposed to individual metals decreased by 50% at 2.67 mg Cd^{2+} kg^{-1} and 5.53 mg Cr^{6+} kg^{-1}. On the other hand addition of magnetite, NPs demonstrated an interactive infra-additive or antagonistic effect against Cd^{2+} and Cr^{6+}, leading to considerable diminishing and toxicity alleviation in vegetable tissue caused by cadmium and chromium accumulation [26].

With the advent of increased and uncontrolled use of NPs, concerns about their possible harmful effects within the environment are also increasing. Several investigations have been conducted to understand their toxic effect on plants as well as the environment in a related study, the effect of TiO and ZnO NPs on wheat growth and soil enzyme activities were estimated under field conditions. Both of the NPs tested reduced the biomass of wheat, whereas TiO_2 NPs were retained in the soil for longer periods by primarily adhering to wheat cell walls. The chances of ZnO NPs getting into wheat was enhanced by its ability to get dissolved in the soil. In addition, the soil enzymes which are bioindicators of soil quality and health, including protease, catalase, and peroxidase activities were inhibited significantly in the presence of NPs. Even though the urease activity was unaffected, the results suggested that NPs themselves or their dissolved ions were clearly toxic for the soil ecosystem [27].

Herbagreen, which involves a tribo-mechanical activation process and which has been successfully exploited for maize cultivation, has also been productively used in wheat. As explained earlier in this chapter, the direct plant penetration property of specialized Herbagreen nanoparticle was also observed in wheat crops. Experiments performed consecutively for two years on wheat have not only shown an increase in crop yield but even show a substantial increase in wheat gluten values. The wheat gluten values in the Herbagreen treated group reached even 29.6%, a value that is well above the FAO recommendation of 20.0%. Herbagreen nanoparticle was even found to challenge the use of chemical fertilizers at their maximum dosage with comparable values of proteins and lipids in the treated plots. It was also concluded that using the Herbagreen fertilizers could avoid or reduce the soil and water contamination caused by only traditional mineral fertilizers use.

KEYWORDS

- **damaged leaf area percentage**
- **glutathione**
- **magnesium oxide nanoparticles**
- **nanoparticles**
- **single cross 704**
- **TiO_2 nanoparticles**

REFERENCES

1. Yang, Z., Jing, C., Dou, R., Gao, X., Mao, C., & Li, W., (2015). Assessment of the phytotoxicity of metal oxide nanoparticles on two crop plants, maize (*Zea mays* L.) and rice (*Oryza sativa* L.), *International Journal of Environmental Research and Public Health, 12*, 15100–15109.
2. Prifti, D., & Maci, A., (2017). Effect of herbagreen nanoparticles on biochemical and technological parameters of cereals (wheat and corn). *European Scientific Journal, 6*, 72–83.
3. Farnia, A., & Omidi, M. M., (2015). Effect of nano-zinc chelate and nano-biofertilizer on yield and yield components of maize (*Zea mays* L.), underwater stress condition. *Indian Journal of Natural, 29*, 4619–4624.
4. Sharifi, R., Mohammadi, K., & Rokhzadi, A., (2016). Effect of seed priming and foliar application with micronutrients on quality of forage corn (*Zea mays*). *Environmental and Experimental Biology, 14*, 151–156.
5. Yaqoob, S., Ullah, F., Mehmood, S., Mahmood, T., Ullah, M., Khattak, A., & Zeb, M. A., (2017). Effect of wastewater treated with TiO nanoparticles on early seedling growth of *Zea mays* L. *Journal of Water Reuse and Desalination* (in press).

6. Hediat, M. H. S., (2012). Effects of silver nanoparticles in some crop plants, common bean (*Phaseolus vulgaris* L.) and corn (*Zea mays* L.). *International Research Journal of Biotechnology*, *10*, 190–197.
7. Jayarambabu, N., Siva, K., B., Venkateswara, R. K., & Prabhu, Y. T., (2016). Enhancement of growth in maize by biogenic-synthesized MgO nanoparticles. *International Journal of Pure and Applied Zoology, 4*, 262–270.
8. Mosanna, R., & Behrozyar, E. K., (2015). Morpho-physiological response of maize (*Zea mays* L.) to zinc nano-chelate foliar and soil application at different growth stages. *Journal on New Biological Reports, 4*, 46–50.
9. Morteza, E., Moaveni, P., Farahani, H. A., & Kiyani, M., (2013). Study of photosynthetic pigments changes of maize (*Zea mays, L.*) under nano TiO spraying at various growth stages. *Springer Plus, 2*, 247.
10. Taheri, M., Qarache, H. A., Qarache, A. A., & Yoosefi, M., (2017). The effects of zinc-oxide nanoparticles on growth parameters of corn (SC704). *STEM Fellowship Journal, Nanoparticles, 1*(2), 17–20.
11. Rico, C. M., Hong, J., Morales, M. I., Zhao, L., Barrios, A. C., Zhang, J. Y., Peralta-Videa, J. R., & Gardea-Torresdey, J. L., (2013). Effect of cerium oxide nanoparticles on rice: A study involving the antioxidant defense system and *in vivo* fluorescence imaging, *Environmental Science & Technology, 47*(11), pp. 5635–5642.
12. Yi, H., Zhang, Z., Rui, Y., Ren, J., Hou, T., Wu, S., Rui, M., Jiang, F., & Liu, L., (2016). Effect of different nanoparticles on seed germination and seedling growth in rice. *2nd Annual International Conference on Advanced Material Engineering*, 166–173.
13. Thuesombat, P., Hannongbua, S., Akasit, S., & Chadchawan, S., (2014). Effect of silver nanoparticles on rice (*Oryza sativa L.* cv. KDML 105) seed germination and seedling growth. *Ecotoxicology and Environmental Safety, 104*, 302–309.
14. Boonyanitipong, P., Kumar, P., Kositsup, B., Baruah, S., & Dutta, J., (2011). Effects of zinc oxide nanoparticles on roots of rice *Oryza sativa* L. *International Conference on Environment and BioScience,* IPCBEEvol.21.
15. Degrassi, G., Antisari, L. V., Venturi, V., Carbone, S., Gatti, A. M., Gambardella, C., Falugi, C., & Vianello, G., (2014). Impact of engineered nanoparticles on virulence of *Xanthomonasoryzae pvoryzae* and on rice sensitivity at its infection. *EQA – Environmental Quality/Qualité de l'Environnement/Qualitàambientale, 16*, 21–33.
16. Elamawi, R. M. A., & El-shafey, R. A. S., (2013). Inhibition effects of silver nanoparticles against rice blast disease caused by Magna-porthe grisea Egypt. *Journal of Agricultural Research, 91*(4).
17. Da Costa. M. V. J., & Sharma, P. K., (2016). Effect of copper oxide nanoparticles on growth, morphology, photosynthesis, and antioxidant response in *Oryza sativa. Photosynthetica, 54*(1), 110–119.
18. Astafurova, T., Zotikova, A., Morgalev, Y., Verkhoturova, G., Postovalova, V., Kulizhskiy, S., & Mikhailova, S., (2015). Effect of platinum nanoparticles on morphological parameters of spring wheat seedlings in a substrate-plant system Nanobiotech. *IOP Publishing IOP Conf. Series: Materials Science and Engineering,* 98.
19. Larue, C., Veronesi, G., Flank, A. M., Surble, S., Herlin-Boime, N., & Carrière, M., (2012). Comparative uptake and impact of TiO2 nanoparticles in wheat and rapeseed. *Journal of Toxicology and Environmental Health, 75*, 722–734.

20. Yasmeen, F., Razzaq, A., Iqbal, M. N., & Jhanzab, H. M., (2015). Effect of silver, copper and iron nanoparticles on wheat germination. *International Journal of Biosciences, 6*, 112–117.
21. Mahmoodzadeh, H., & Aghili, R., (2014). Effect on germination and early growth characteristics in wheat plants (*Triticum aestivum,* L.) seeds exposed to TiO. *Journal of Chemical Health Risks, 4*(1), 29–36.
22. Barbasz, A. B., & Ocwieja, K. M., (2016). Effects of exposure of callus cells of two wheat varieties to silver nanoparticles and silver salt ($AgNO_3$). *Acta Physiologiae Plantarum, 38*, 76.
23. Christian, O., Dimkpa, J. E. McLean, N. M., David, W. B., Richard, H., & Anne, J. A., (2013). Silver nanoparticles disrupt wheat (*Triticum aestivum,* L.) growth in a sand matrix environ. *Science and Technology*, *47*(2), 1082–1090.
24. Iannone, M. F., Groppa, M. D., De Sousa, E. M., Beatriz, M., Raap, F., & Benavides, M. P., (2016). Impact of magnetite iron oxide nanoparticles on wheat (*Triticum aestivum* L.) development: Evaluation of oxidative damage. *Environmental and Experimental Botany*, *131*, 77–88.
25. Rafique, R., Arshad, M., Khokhar, M. F., Qazi, I. A., Hamza, A., & Virk, N., (2014). Growth response of wheat to Titania nanoparticles application. *NUST Journal of Engineering Sciences*, *7*, 42–46.
26. López-Luna, J., Silva-Silva, M. J., Martinez-Vargas, S., Mijangos-Ricardez, O. F., GonzálezChávez, M. C., Solís-Domíngue, F. A., & Cuevas-Díaz, M. C., (2016). Magnetite nanoparticle (NP) uptake by wheat plants and its effect on cadmium and chromium toxicological behavior. *Science of the Total Environment*, *565*, 941–950.
27. Du, W., Sun, Y. R., Wu, J., & Guo, H., (2011). iTiO_2 and ZnO nanoparticles negatively affect wheat growth and soil enzyme activities in agricultural soil. *J. Environ. Monit.*, *13*, 822–828.
28. Datta, S. K., (2004). Rice biotechnology: A need for developing countries. *Journal of Agrobiotechnology Management and Science*, *7*(1&2), Article 6.

CHAPTER 7

Nanotoxicology: A Threat to Human Health and Environment

SUSHMITA SHRIVASTAVA* and RAJESH SINGH TOMAR

Amity Institute of Biotechnology, Amity University Madhya Pradesh, Gwalior, India

**Corresponding author. E-mail: rssush@gmail.com*

ABSTRACT

Nanotoxicology is a particle toxicity deals with the adverse effects of nanoparticles (NPs) on human health and the environment. Particles of 100 nm or less are considered as challenging under toxicological studies. They impose oxygen stress and activate redox sensitivity in cell membranes and cell organelles, and their assays are responsible for genotoxicity. NPs have a high surface area to unit mass ratio, which leads to inflammatory effects on the organs and can translocate via blood to different parts of the body, causing abnormalities in blood, brain, liver, skin, and gut. They may induce cytokine production and cause cell death. Metal-based NPs have a negative impact on the environment. Coating and charges on NPs also impose toxicity. They can accumulate in the food chain and transferred to various trophic levels producing stronger toxicity.

7.1 INTRODUCTION

Nanotechnology is a technique in which a very small particle about 100 nm or less is utilized for various productive purposes. They can be produced by any chemical. It is beneficial for the production of energy, food, solar cells, electronic devices, etc. It provides huge benefits for society, but simultaneously, it causes adverse effects on human health and the environment. Such adverse effects are considered under nanotoxicology. Fine nanoparticles (NPs) are found to be more toxic as compared to large ones. Thus nanotoxicology is the science which deals which the effects and capability of NPs to harm the living system. Nanomaterials are generally the part of industrial products, which are produced by synthetic or manufacturing processes or maybe originated by combustion or simply by nature.

7.1.1 ROUTE OF TOXICITY

NPs can be entered into the body by several means, either by dermal exposure or in inhalation or by oral intake [1]. They can easily cross the biological barrier due to their small size. They are not recognized by the defense mechanism of our body. They can even reach the brain via blood, affecting the central nervous system. Nanotoxicity depends on the dose, surface area, size, structure, coating, and opsonization [2]. Sunscreens have Tio_2 or Zno_2, which are susceptible to the formation of ROs. NPs react with an enzymatic pathway, which hampers the degradation of Phenylalanine [3].

7.1.2 EFFECT ON HUMANS

NPs can enter into a man by air, water, and skin. They can then enter into internal tissues and causing damage. NPs impose oxygen stress and cause redox sensitivity in cell membranes and other organelles like mitochondria, nucleus, etc. They lead to the production of oxygen reactive species, which also damage protein and even DNA. They enhance cytokine production and cause cell death (apoptosis and necrosis).

Nanomaterials can enter into the skin via cosmetics and may accelerate acne, eczema, wounds, sunburns, etc. Carbon nanotubes (CNTs) cause mesothelioma (pleural abnormality). Metal oxides like titanium oxide, zinc oxide, carbon black, and CNTs lead to inflammation of lungs and may translocate to various other organs through blood [4]. Citrate-capped gold nanoparticles (AuNPs) are found to be toxic to human carcinoma lung cell lines.

Some metallic oxides also lead to genotoxicity. Even assays like comet assay, micronucleus assay, chromosomal aberration test are also responsible for causing genotoxicity. Copper NPs are used in lipsticks, can enter into the digestive tract, and cause metabolic alkalosis. SiO_2 causes inflammation in the submucosa cells of the lungs. It also causes a decrease in mitochondrial activity. NPs can also enter into the embryo through the placenta [5].

7.1.3 EFFECT ON ANIMALS

It has been reported that CNTs cause bovine spongiform encephalopathy (Mad's cow disease). NPs affect fish physiology; they killed water fleas, affect soft-bodied animals, may accumulate in gills, intestine, testis, liver, the brain of Medaka (*Oryzias latipes*) fish [6]. Silica and titanium oxide, when injected in mice, crossed the placenta and caused severe complications in pregnancy and may affect their offspring. NPs can enter into blood vessels of the chicken embryo. It has been reported that gold particles are used to study size-dependent biological response, but when 8–37 nm NPs are injected into mice, they elicit severe sickness. Similarly, Fe also affects nerve growth factors. Metal-based NPs induce toxicity in mammals by interacting with proteins and enzymes. Aluminum oxide causes genotoxicity in rats. Single-walled nanotubes at high concentrations produce granulomas in the lungs of rats.

7.1.4 EFFECT ON PLANTS

Plants are also affected by NPs, particularly metal-based. Nanotubes cause an increase in hydrogen peroxide products, which relates to oxidative stress and cell death. NPs may cause blockage in intercellular spaces. ZnO affects seed germination. Single-walled carbon nanotubes (SWCNTs) affect root elongation in certain plants like tomato, lettuce, carrot, cabbage, etc. They also affect soil microbial activity and diversity. Even ZnO could absorb into ryegrass root causing damage to root cells and inhibits seedling growth [7].

7.1.5 EFFECT ON MICROORGANISMS

Several microbiologists have done enormous work on the toxicity of NPs on microbes and found that ZnO is highly toxic, followed by CuO, TiO_2, etc.

7.1.6 EFFECT ON ENVIRONMENT

NPs accumulate in environmental and biological fluids due to high ionic strength. Metal-based NPs are utilized in the drug delivery system have a negative impact on the environment. Charges and coatings on particles also affect toxicity [8]. It has been found that polyvinyl pyrrolidone (PVP) coated silver nanoparticles (AgNPs) gets transferred from *E. coli* to *Caenorhabditis elegans* via the food chain and gets accumulated in the gut, subcutaneous tissue and gonad causing germ cell death and affect several generations. Small AgNPs (25 nm) get easily accumulated in the food chain and exhibited stronger toxicity to higher trophic levels.

7.2 CONCLUSION

No doubt that nanotechnology has an enormous scope and potential for the betterment of society, but still, their harmful aspects need to be noticeable, which affects each and every trophic level of the ecosystem. There should be limited and safe use of these particles. People must be aware of the harmful aspects of NPs. Even some regulatory acts are also enacted in the US for controlling the sale and use of new chemical substances, which may affect public health and the environment [9]. Similarly, in Europe, REACH (Registration, Evolution, Authorization, and Restriction of Chemical Substances) is formed. Regulations are also formed for food and cosmetics.

KEYWORDS

- **genotoxicity**
- **nanoparticles**
- **nanotoxicology**
- **polyvinyl pyrrolidone**
- **silver nanoparticles**
- **single-walled carbon nanotubes**

REFERENCES

1. Ehrhart, F., Evelo, C. T., & Willighagen, E., (2015). Current systems biology approaches in hazard assessment of nanoparticles. *Nanotoxicology*. doi: http://dx.doi.org/10.1101/028811 (Accessed on 5 November 2019).
2. Donaldson, K., Stone, V., Trans, C. L., Kreyling, W., & Borm, P. J. A., (2004). Nanotoxicology. *Toxicology*, 727–728.
3. Krug, H. F., & Wick, P., (2011). Nanotoxicology: An interdisciplinary challenge. *Nanotoxicology*, 50. doi: 10.002/ anie.201001037, 1261–1280.
4. Bacanli, M., & Basaran, N., (2014). Nanotoxicology: New area in toxicology. *Turk, J. Pharm. Sci., 11*(2), 231–240.
5. Seabra, A. B., & Duran, N., (2015). Nanotoxicology of metal oxide nanoparticles. *Metals, 5*, 934–975. doi: 10.3390/met5020934.
6. Rajkishore, S. K., Subramanian, K. S. N., & Gunasekharan, K., (2013). Nanotoxicity at various trophic levels: A review. *Bioscan, 8*(3), 975–982.
7. Battacharya, A., Mohammad, F., Naika, H. R., Epid, S. T. T., Reddy, J. R., & Prakashan, R. S., (2015). Nanoparticles and their impact on plants. *Research Journal of Nanoscience and Nanotechnology, 5*(2), 27–43.
8. Farhan, M., Khan, I., & Thiagarajan, P., (2014). Nanotoxicology and its implications. *Research Journal of Pharmaceutical, Biological and Chemical Sciences, 5*(1), 470–479.
9. Drobne, D., (2007). Nanotoxicology for safe and sustainable nanotechnology, *Arh. Hig. Rada. Toksikol., 58*, 471–478.

CHAPTER 8

Impact of Nanotoxicity on Human Health: An Overview

BHARTI PRAKASH* and S. MATHUR

Department of Zoology, SPC Government College, Ajmer, Rajasthan, India

**Corresponding author. E-mail: dr.bharti.prakash@gmail.com*

ABSTRACT

Nanotechnology has an important part in our daily life. Nanomaterials are used in different fields like textiles, medical devices, drugs, cosmetics, and various other things. Nanoparticles (NPs) penetrate into the human body through various channels. The size and surface of NPs play an important role in determining the toxicity. In this chapter, we tried to find out the adverse influence of NPs on human health and the environment.

8.1 INTRODUCTION

Over the last decade, a wide variety of nanomaterials have been manufactured and used in different products. The development of nanotechnology in different industries, its modernity, and also the lack of information on its negative effects on human health and the environment originate from the novel mechanisms that are also related to nanotoxicology. Human and environmental exposures are inevitable, and the more widespread these NPs enabled products to become, the bigger is the concern around their safety and impact on human and environmental health.

8.1.1 *WHAT IS NANOTECHNOLOGY?*

"Nanotechnology" refers to the design, production, and application of structures, devices, or systems at the incredibly small scale of atoms and molecules – the "nanoscale." "Nanoscience" is the study of phenomena and the manipulation of materials at this scale, generally understood to be 100 nanometers (nm) or less [64, 87]. The prefix "nano" means one-billionth or 10^{-9}; the word itself comes from the Greek word Nanos, meaning dwarf. Hence one nanometer is equivalent to one billionth of a meter. Things on the nanoscale are 100–1,000 times smaller than anything that can be seen with an optical microscope.

NPs, an important subset of nanomaterials, are defined as discrete particles with at least one dimension less than 100 nm (which includes thin enough fibers). NPs can have various forms and shapes and include quantum dots (QDs), fullerenes, nanotubes, and nanowires [59].

The nanomaterials are classified into three categories, depending on the location of the nanoscale structure in the system (Figure 8.1) [102].

1. Materials that are nanostructured in bulk;
2. Materials that have nanostructure on the surface; and
3. Materials that contain nanostructured particles. This is further classified into four subcategories which can be identified depending on the environment around the NPs:

   i. **Subcategory III a:** Nanoparticles bound to the surface of another solid structure.
   ii. **Subcategory III b:** Systems where nanoparticles are suspended in a liquid.
   iii. **Subcategory III c:** Nanoparticles suspended in solids.
   iv. **Subcategory III d:** of airborne nanoparticles.

8.2 OBJECTIVES

The database studied so far shows the research gap in the field of NPs and its toxicity. The present paper tries to explore the toxicity of nanomaterials on the human body and also tries to find out the benefits of nanotechnology-based products and its effect on humans.

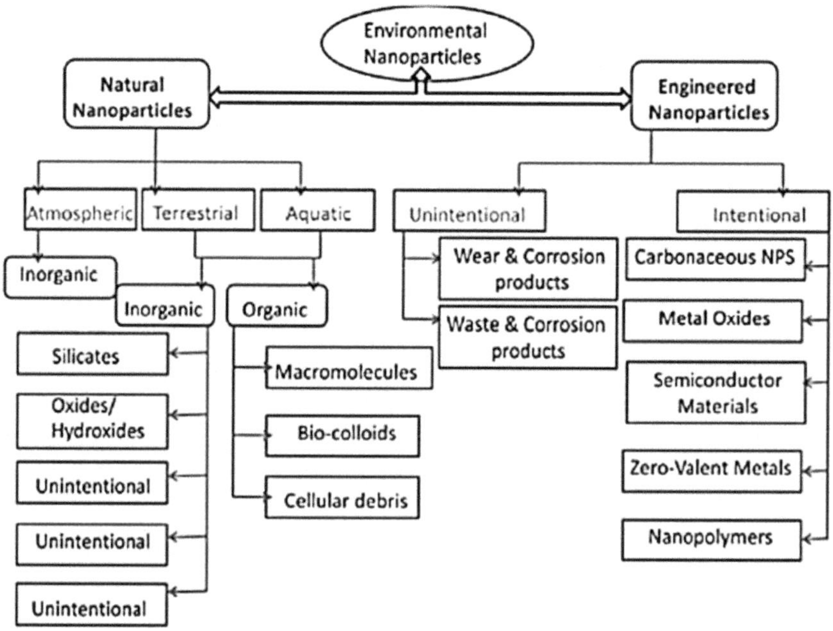

FIGURE 8.1 Categorization of nanoparticles present in the environments [102].

8.3 METHODOLOGY

In the present paper, different scientific literature, articles of peer-reviewed journals, and search engines on the toxicity of nanomaterials and their effect on the environment and mainly human health have been analyzed and summarized.

8.3.1 NANOTOXICOLOGICAL STUDIES

Nanotechnology has wide applications in many fields, especially in the biological sciences and medicine. Nanomaterials are applied as coating materials or in treatment and diagnosis. According to the WHO report, two cases of intoxication of NPs and their effects on human health have been reported. In the first case, breathlessness and excess of liquid in the lungs, and death has been reported in Chinese women due to polyacrylate NPs exposure during January–April 2008 [58, 98]. In another case in late

March 2006, a similar symptom of breathlessness and pulmonary edema was reported in Germany [15, 31, 56, 80]. From the several studies carried out on various forms of nanomaterials (C60, single-, and multiwalled carbon nanotubes (CNTs), among others) it is observed that the toxicological response was related to "dose by mass," i.e., the higher the dose that laboratory animals are exposed to, the more severe the adverse effect observed [65].

The extensive application of nanomaterials in a wide range of products for human use poses a potential for toxicity risk to human health and the environment. Such adverse effects of nanomaterials on human health have triggered the development of a new scientific discipline known as "nanotoxicity"-the study of the toxicity of nanomaterials. The manmade nanoparticle (NP) in recent times have found entry into consumer goods like cosmetics, medical devices, household appliances, and even food packaging [96]. Such extensive application of nanomaterials in different consumer products poses a potential toxicity risk to human health and the environment [2–4, 56, 66, 68, 83, 88].

The size, shape, charge, surface, and other factors of NPs show their effect on human health. However, several reviews of the toxicology of NPs underscore the importance, as well as the unique difficulties of nanotoxicology studies [10, 12, 21, 23, 26, 29, 36, 50, 72, 74, 100]. The properties of nanomaterials that influence toxicity also include chemical composition, aggregation, and solubility, and the presence of "functional groups" of other chemicals [54, 68].

Biological effects caused by nanoparticle deposition are related to their physical and chemical parameters [18, 28, 32, 42, 48, 73]. These properties of NPs have a significant role in influencing movement, adhesion, and modulation of physiological pathways [32, 37, 67, 68].

For proper characterization of NPs for toxicological studies, there are 15 following parameters:

1. Mass, concentration;
2. Chemical composition (purity and impurities);
3. Solubility;
4. Specific area;
5. Number of particles;
6. Particle size and distribution;

7. Surface properties (charge/zeta potential, reactivity, chemical composition, functional Groups, redox potential, potential to generate free radicals, presence of metals, surface covering, etc.);
8. Shape, porosity;
9. Degree of agglomeration/aggregation;
10. Biopersistence;
11. Crystalline structure;
12. Hydrophilicity/hydrophobicity;
13. Site of pulmonary deposition;
14. Age of particles; and
15. Producer, process, and source of material used.

8.3.2 *WHAT ARE POTENTIAL HARMFUL EFFECTS OF NANOPARTICLES (NPS)?*

The widespread use of NP in agricultural and industrial products result into these particles being released into the environment by waste disposal and industrial sewage. Once they enter the environment, NP ultimately enters the food chain because these particles are absorbed by different bio-organisms [25, 92]. The distress over the possible unfavorable impact of NP on the biological systems has given rise to nanotoxicology. Many activist groups, as well researchers, are anxious and concerned about nanotoxicity to humans and the environment [16, 39, 84].

Various epidemiological studies have been associated with the exposure of combustion-generated fine particles with lung cancer, heart disease, asthma, and/or increased mortality. Epidemiological proof links cardio-respiratory endpoints to outdoor air pollution. NPs, even though they are very small, get deposited not only in the alveolar region but in the entire pulmonary system. NPs can enter the body quite easily through dermal absorption, inhalation, and gastrointestinal (GI) assimilation. They can also enter the central nervous system through the olfactory mucosa or through the blood-brain barrier, especially during the fetal stage and early life. These properties are of great interest for the targeted delivery of drugs. Many NPs intended for medical imaging and/or therapeutics are still in an investigational stage. Importantly, nanomedical products can expose workers as well as patients, so these are not entirely distinct kinds of exposures [17, 20, 41, 43, 55, 63, 75, 76, 90, 91, 107].

Studies have also reported the cytotoxic effects of nanospheres, nanoshells, and nanocapsules on the cell membrane, mitochondrial function, prooxidant/antioxidant status, enzyme leakage, DNA, and other biochemical endpoints. These accounts are also explained by other researchers (Figure 8.2) [27, 40, 51, 79, 102].

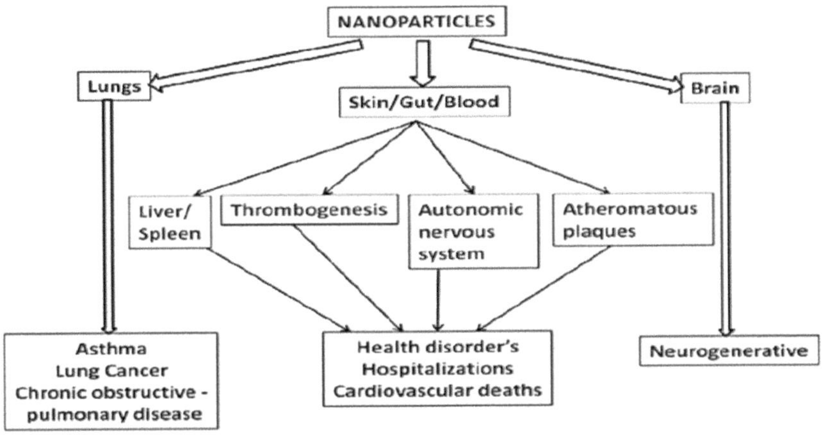

FIGURE 8.2 Systematic health effects of nanoparticles on the human body [102].

In fact, for workers who handle nanomaterials prepared and used in colloidal forms, cutaneous absorption is an important route of exposure [19]. Liposoluble NPs can possibly move in the intercellular space of the corneal layer and make its way through the cells, the hair follicles, and the sweat glands. The unvascularized zone of the skin can also act as storage for NPs, from where they cannot be eliminated by macrophages [62].

Current studies show that the heart might be the target organ for carbon NP toxicity. New techniques and health assessment methods are required to be developed for recognizing sub-optimal health in populations exposed to carbon NPs. Various procedures by which NPs can cause cell damage have been investigated in both chemical (e.g., oxidative stress, inflammation, protein misfolding) and physical nature (e.g., direct physical damage, production of secondary photoelectrons), involving the cell nucleus, the membrane, and cell organelles. Some particles can be moved along the sensory axons to the central nervous system [61, 73, 86, 103].

From the toxicological point of view, CNTs are fibers, and their toxicity is correlated to their persistence in the lungs [49]. The studies of displacement of 14 nm carbon black particles by tracheal installation, shows nanoparticle buildup in the spaces between the alveolar epithelial cells with passage into the bloodstream, perhaps involving cytoplasm shrinkage due to stimuli generated by nanoparticle-mediated bonding with the alveolar epithelial cells, followed by sporadic NP dissemination of the alveolar membrane [97]. The chief routes of occupational and environmental NP exposures are through the skin, GI tract, and respiratory tract [5, 20, 33, 43, 60, 99].

Fullerenes are spherical cages containing from 28 to more than 100 carbon atoms. The most widely studied form containing 60 carbon atoms, C60, was synthesized first time in 1985 [46]. Among the many biomedical applications proposed for fullerenes, it is exhibited that fullerenes can be excellent sensors of free radicals [47, 71]. Various epidemiological studies showed a direct relationship between exposure to nano-scaled ultrafine dusts and respiratory and cardiovascular effects. Notable relationships are established in several epidemiological studies demonstrating an increase in fine particles in air pollution, from vehicle emissions, leading to an increase in morbidity and mortality of fragile populations with respiratory and cardiac problems [1, 7, 13, 14, 24, 25, 45, 53, 70, 77, 78, 81, 82, 85, 89, 93–95, 101, 104–106]. Ultrafine dusts such as welding fumes, diesel emissions, etc. are unfavorable secondary reaction products. However, NPs are essentially new-engineered particles. Their production depends on their unique properties, based on their small size, their large specific surface, and the quantal effects, which allow consideration of new industrial and commercial perspectives. The ultrafine particles, which have granulometric properties similar to engineered NPs, show toxic effects of different natures in many organs, even if they are absorbed almost exclusively by the pulmonary route. Their effects have been reported in animals, and their effects are also shown by different clinical and epidemiological studies in humans. Particles deposited in the respiratory system that are cleared via the mucociliary escalator may be swallowed, leading to exposure to the GI tract. Additional ingestion routes include the use of nanostructured materials in food, water, and drugs. Relatively few studies have investigated nanostructured materials in the GI tract, and most have shown them to pass through and be eliminated rapidly (Figure 8.3) [17, 56, 57].

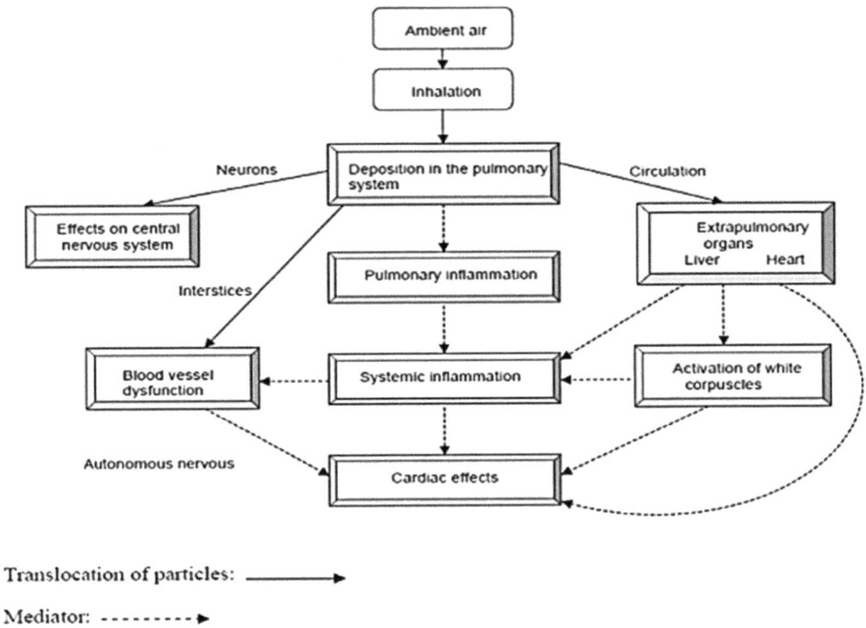

FIGURE 8.3 Potential effects of inhaled ultrafine particles [17, 72] (Reproduced with permission of Dr. Gunter Oberdörster, 2005).

A recent study showed that exposing the mesothelial lining of the body cavity of mice to long MWCNT results in asbestos-like pathogenic behavior, including inflammation and formation of granulomas [83]. Cutaneous studies of TiO_2 in various sunscreen formulations did not show absorption beyond the dermis (in healthy skin) in human subjects. Dermal exposure to NPs also occurs with consumer products. Current NP exposures in consumer products include inorganic NPs. Consumers are exposed to TiO_2 and ZnO NPs in sunscreens [60]. In addition, wound dressings have been reported to contain nanocrystalline silver as an antimicrobial agent [8, 9]. The cytotoxicity of ZnS and CdSe nanocrystal solutions for human fibroblasts and tumor cells [44]. Cytotoxicity was higher if the nanocrystals were coated with mercaptopropionic acid, an unstable coating. Phosphosilicate, PEG-silica, and polymer-coated inert gold nanoparticles (AuNPs) also have a cytotoxic effect [11, 30, 35, 38, 68].

8.4 CONCLUSION

A control level on manufacturing and commerce of nanoproducts should be exercised by regulatory bodies. Manufacturers must be fully aware of the nanomaterials present in their products and should get a license for manufacturing the same.

1. There should be a chemical abstracts service registry number (CAS RN) for nanoscale material.
2. Mandatory monitoring schemes and health surveillance systems should be instituted by Governments and international agencies.
3. A constant monitoring of the workers for the diseases relevant to the potential exposure should be taken up with priority.
4. Green nanoscience should be used to develop safer and green substitutes in place of lethal nanomaterial.
5. Green nanoscience can also diminish the hazards of nanomaterial production and increase its potentiality.
6. It is recommended that the consideration of ethical and social implications of nanotechnologies should form part of the formal training of all research students and staff working in these areas

The goal must be to realize the great opportunities and benefits of nanotechnology while at the same time minimizing the risk related to its applications. The use of nanomaterials to achieve the goals of green chemistry will provide new opportunities to build up superior products and processes with reduced adverse impact on the environment and human health (Table 8.1).

TABLE 8.1 List of Some Existing ENPs and Their Health and Environmental Effects [102]

Nanoparticle	Environmental Effects	Health Effects
Carbon nanotubes	Cause indirect effects upon contact with the surface of the environmental organisms; environmental damage	Apoptosis; decreased cell viability; lung toxicity; oxidative stress; retarded cell growth; skin irritation, etc.
Fullerenes	Effects on soil organisms and enzymes; aquatic ecosystems; binding of chemicals to fullerenes may affect the toxicity of other environmental contaminants	Retarded cell growth; decreased cell viability; oxidative stress and apoptosis, etc.

TABLE 8.1 *(Continued)*

Nanoparticle	Environmental Effects	Health Effects
Heterogeneous nanostructures	Toxicity depends on multiple physicochemical as well as environmental factors; the adverse influence of the ecosystem, etc.	Arrest of cell growth and sometimes even cell death; chromatin condensation; free radical formation
Nanosilver	Undergoes several transformations when it is released into the environment and shows adverse effects	Alterations of the non-specific immune responses; altered cell signaling; apoptosis; necrosis of cells; oxidative stress, etc.
Nano-structured flame retardants	Persistent and tend to accumulate in the environment, toxic to plants, wildlife, etc.	Oxidative stress; fibrosis; cardiovascular effects; cytotoxicity; carcinogenic, etc.
Polymeric nanoparticle	Potential hazardous factor for environmental exposure	Oxidative stress; inflammation; alteration in cellular morphology and functioning, etc.
Silicon based nanoparticles	Potential hazardous factor for environmental exposure; adverse influence of ecosystem, etc.	Cardiovascular effects; cytotoxicity, increase oxidative stress, etc.
TiO_2 nanoparticles	Disrupt aquatic ecosystems carbon and nitrogen cycles; stress photosynthetic organisms	Excessive exposure in humans may result in increased oxidative stress, retarded cell growth, slight changes in the lungs, etc.

KEYWORDS

- **carbon nanotubes**
- **chemical abstracts service registry number**
- **nanoparticles**
- **nanotechnology**
- **nanotoxicity**
- **quantum dots**

REFERENCES

1. Ai, J., Biazar, E., Jafarpour, M., Montazeri, M., Majdi, A., Aminifard, S., Zafari, M., Akbari, H. R., & Rad, H., (2011). Nanotoxicology and nanoparticle safety in biomedical designs. *Int. J. Nanomedicine*, *6*, 1117–1127. doi: 10.2147/IJN.S16603.
2. Sharma, M. (2010). Understanding the mechanism of toxicity of carbon nanoparticles in humans in the new millennium: A systemic review. *Indian Journal of Occupational and Environmental Medicine, 14*(1), January–April, 2010, pp. 3–5. *Archived DTSC Nanotechnology Symposia*. Department of Toxic Substances Control. https://dtsc.ca.gov/TechnologyDevelopment/Nanotechnology/nanoport.cfm.
3. *An Issues Landscape for Nanotechnology Standards*, (2007). Report of a Workshop. Institute for Food and Agricultural Standards, Michigan State University, East Lansing.
4. Pollution particles lead to higher heart attack risk (update). Available from: http://www.bloomberg.com/apps/newsstudy (Accessed on 5 November 2019).
5. Aitken, R. J., Chaudhry, M. Q., Boxall, A. B., & Hull, M., (2006). Manufacture and use of nanomaterials: Current status in the UK and global trends. *Occup. Med. (Lond.), 56*, 3006. Google Scholar Medline.
6. Balbus, J. M., Castranova, V., Colvin, V. L., Daston, G. P., Denison, R. A., & Maynard, A. D., (2007). Meeting report: Hazard assessment for nanoparticles—Report from an interdisciplinary workshop. *Environmental Health Perspectives, 115*, 1654–1660.
7. Bateson, T. F., & Schwartz, J., (2001). Selection bias and confounding in case-crossover analyses of environmental time-series data. *Epidemiology, 12*, 654–661.
8. Bhattacharyya, M., & Bradley, H., (2006). Management of a difficult-to-heal chronic wound infected with methycillin-resistant *Staphylococcus aureus* in a patient with psoriasis following a complex knee surgery. *Int. J. Low Extrem. Wounds, 5*, 105–108. Google Scholar Abstract.
9. Bhattacharyya, M., & Bradley, H., (2008). A case report of the use of nanocrystalline silver dressing in the management of acute surgical site wound infected with MRSA to prevent cutaneous necrosis following revision surgery. *Int. J. Low Extrem. Wounds, 7*, 45–48. Google Scholar Abstract.
10. Bonner, J. C., (2010). Nanoparticles as a potential cause of pleural and interstitial lung disease. *Proc. Am. Thorac. Soc., 7*, 138–141. Google Scholar Medline.
11. Borm, P. J. A., (2002). Particle toxicology: From coal mining to nanotechnology. *Inhalation Toxicology, 14*, 311–324.
12. Boverhof, D. R., & David, R. M., (2010). Nanomaterial characterization: Considerations and needs for hazard assessment and safety evaluation. *Anal. Bioanal. Chem., 396*, 953–961. Google Scholar Medline.
13. Brunekreef, B., Janssen, N. A. H., De Hartog, J. J., Oldenwening, M., Meliefste, K., Hoek, G., Lanki, T., Timonen, K. L., Vallius, M., Pekkanen, J., & Van Grieken, R., (2005). Personal indoor and outdoor exposures to PM2.5 and its components for groups of cardiovascular patients in Amsterdam and Helsinki. *Health Effects Institute Report, 127*, p. 79.
14. Bruske-Hohlfeld, I., Peters, A., & Wichmann, H. E., (2005). Epidemiology of nanoparticles. In: Edited by the Health and Safety Executive, *Proceedings of the First International Symposium on Occupational Health Implications of Nanomaterials*

(pp. 53–58). Buxton, Great-Britain. Great-Britain and the National Institute for Occupational Safety and Health, United States. http://www.hsl.gov.uk/capabilities/nanosymrep_final.pdf (Accessed on 5 November 2019).
15. Bulls, K., (2006). *"Nano" Safety Recall (Web Site)*. Cambridge, M. A., MIT technology review. http://www.technologyreview.com/NanoTech/wtr_16681,318,p1.html (Accessed on 5 November 2019).
16. Buzea, C., Pacheco, I., & Robbie, K., (2007). Nanomaterials and nanoparticles: Sources and toxicity. *Bio Interphases, 2*, MR17–MR71 (PubMed).
17. Claude, O., Soucy, B., Lapointe, G., Woods, C., & Ménard, L., (2008). *Health Effects of Nanoparticles Second Edition Chemical Substances and Biological Agents Report r-589 Studies and Research Projects IRSST – Communications Division 505.* Demaisonneuve blvd west montréal (québec) H3A 3C2 publications@irsst.qc.ca. 4–8 pp. www.irsst.qc.ca.
18. Cohen, B. S., Li, W., Xiong, J. Q., & Lippmann, M., (2000). Detecting H+ in ultrafine ambient aerosol using iron nano-film detectors and scanning probe microscopy. *Applied Occupational and Environmental Hygiene, 15*, 80–89.
19. Colvin, V. L., (2003). The potential environmental impact of engineered nanomaterials. *Nature Biotechnology, 21*(10), 1166–1170.
20. Cormode, D. P., Skajaa, T., Fayad, Z. A., & Mulder, W. J., (2009). Nanotechnology in medical imaging: Probe design and applications. *Arterioscler. Thromb. Vasc. Biol., 29*, 992–1000. Google Scholar Medline.
21. De Jong, W. H., & Borm, P. J., (2008). Drug delivery and nanoparticles: Applications and hazards. *Int. J. Nanomedicine, 3*, 133–149. Google Scholar Medline.
22. Dhawan, A., Sharma, V., & Parmar, D., (2009). Nanomaterials: A challenge for toxicologists. *Nanotoxicology, 3*, 1–9.
23. Doak, S. H., Griffiths, S. M., Manshian, B., Singh, N., Williams, P. M., Brown, A. P., & Jenkins, G. J., (2009). Confounding experimental considerations in nanogenotoxicology. *Mutagenesis, 24*, 285–293. Google Scholar Medline.
24. Dockery, D. W., & Pope, C. A., (1994). *Annu. Rev. Public Health, 15*, 107.
25. Dominici, F., Peng, R. D., Bell, M. L., Pham, L., McDermott, A., Zeger, S. L., & Samet, J. M., (2006). Fine particulate air pollution and hospital admission for cardiovascular and respiratory diseases. *JAMA, 295*(10), 1127–1134.
26. Duffin, R., Mills, N. L., & Donaldson, K., (2007). Nanoparticles—a thoracic toxicology perspective. *Yonsei. Med. J., 48*, 561–572. Google Scholar Medline.
27. El-Ansary, A., Al-Daihan, S., Bacha, A. B., & Kotb, M., (2015). Toxicity of novel nanosized formulations used in medicine. *Methods Mol. Biol., 1028*, 47–74.
28. Frampton, M. W., (2000). *Systemic and Cardiovascular Effects of Airway Injury and Inflammation: Ultrafine Particle Exposure in Humans* (pp. 529–532). Presented at the workshop in inhaled environmental/occupational irritants and allergens: Mechanisms of cardiovascular and systemic responses, Scottsdale.
29. Fischer, H. C., & Chan, W. C., (2007). Nanotoxicity: The growing need for *in vivo* study. *Curr. Opin. Biotechnol., 18*, 565–571. Google Scholar Medline.
30. Gambardella, C., Morgana, S., Bari, G. D., Ramoino, P., Bramini, M., Diaspro, A. Falugi, C., & Faimali, M., (2015). Multidisciplinary screening of toxicity induced by silica nanoparticles during sea urchin development. *Chemosphere, 139*, 486–495.

31. Glaza, P., (2010). *Industry Should Take Lead with Media When an Event Like a Recall Occurs.* Electronic (web site). http://www.electroiq.com/articles/stm/print/volume-6/issue-3/opinion/guest-column/industry-should-take-leadwith-media-when-an-event-like-a-recall-occurs.html (Accessed on 5 November 2019).
32. Hamoir, J., Gustin, P., Halloy, D., Nemmar, A., Nemery, B., Vanderplasschen, A., Vincke, G., & Wirth, D., (2003). Effect of polystyrene particles on lung microvascular permeability in isolated perfused rabbit lungs: Role of sizeand surface properties. *Toxicology and Applied Pharmacology, 190,* 278–285. http://dx.doi.org/10.1016/s0041–008x[03]00192–3 (Accessed on 5 November 2019).
33. Han, J. H., Lee, E. J., Lee, J. H., So, K. P., Lee, Y. H., Bae, G. N., Lee, S. B., Ji, J. H., Cho, M. H., & Yu, I. J., (2008). Monitoring multiwalled carbon nanotube exposure in carbon nanotube research facility. *Inhal. Toxicol., 20,* 741–749. Google Scholar Medline.
34. Hansen, S. F., Larsen, B. H., Olsen, S. I., & Baun, A., (2007). Categorization framework to aid hazard identification of nanomaterials. *Nanotoxicology, 1,* 243–250.
35. Henneberger, A., Cyrys, J., Couderc, J. P., Ibald-Mulli, A., Rückerl, R., & Wojciech, Z., (2005). Repolarization changes induced by air pollution in ischemic heart disease patients. *Environmental Health Perspectives, 113,* 440–446.
36. Hoet, P., Legiest, B., Geys, J., & Nemery, B., (2009). Do nanomedicines require novel safety assessments to ensure their safety for long-term human use? *Drug Saf., 32,* 625–636. Google Scholar Medline.
37. Hubbs, A. F., Mercer, R. R., Benkovic, S. A., Harkema, J., Sriram, K., Schwegler-Berry, D., Goravanahally, M. P., Nurkiewicz, T. R., Castranova, V., & Sargent, L. M., (2011). *Nanotoxicology – A Pathologist's Perspective Toxicologic Pathology, 39*(2), 2011.
38. Hussain, S. M., Braydich-Stolle, L. K., Schrand, A. M., Murdock, R. C., Yu, K. O., Mattie, D. M., Schlager, J. J., & Terrones, M., (2009). Toxicity evaluation for safe use of nanomaterials: Recent achievements and technical challenges. *Advanced Materials, 21,* 1549–1559. http://dx.doi.org/10.1002/adma.200801395 (Accessed on 5 November 2019).
39. Hutchison, J. E., (2008). Greener nanoscience: A proactive approach to advancing applications and reducing implications of nanotechnology. *ACS Nano, 2,* 395–402.
40. Jeannet, N., Fierz, M., Schneider, S., Künzi, L., Baumlin, N., Salathe, M., Burtscher, H., & Geiser, M., (2015). Acute toxicity of silver and carbon nanoaerosols to normal and cystic fibrosis human bronchial epithelial cells. *Nanotoxicology, 26,* 1–13.
41. KAS, (2009). Are nano EHS budgets too small? *Environmental Health, Occupational Health and Safety, the Pump Handle.* https://thepumphandle.wordpress.com/2009/06/01/are-nano-ehs-budgets-too-small/ (Accessed on 5 November 2019).
42. Kreyling, W. G., Erbe, F., Mayer, P., Oberdörster, G., Semmler, M., Schultz, H., & Takenaka, S., (2002). Translocation of ultrafine insoluble iridium particles from lung epithelium to extrapulmonary organs is size dependant but very low. *Journal of Toxicology and Environmental Health, 65,* 1513–1530.
43. Kim, Y. S., Kim, J. S., Cho, H. S., Rha, D. S., Kim, J. M., Park, J. D., et al., (2008). Twenty-eight-day oral toxicity, genotoxicity, and gender-related tissue distribution of silver nanoparticles in Sprague-Dawley rats. *Inhal. Toxicol., 20,* 575–583. Google Scholar Medline.

44. Kirchner, C., Liedl, T., Kudera, S., Pellegrino, T., Munoz Javier, A., Gaub, H. E., Stolzle, S., Fertig, N., & Parak, W. J., (2005). Cytotoxicity of colloidal CdSe and CdSe/ZnS nanoparticles. *Nano Lett., 5*(2), 331–338.
45. Kreyling, W. G., Semmler, M., & Moller, W., (2004). Dosimetry and toxicology of ultrafine particles. *Journal of Aerosol Medicine, 17*(2), 140–152.
46. Kroto, H. W., Heath, J. R., & O'Brian, S. C., (1985). C60: Buckminsterfullerene. *Nature, 318*, 162–163.
47. Krusic, P. J., Wasserman, E., Keizer, P. N., Morton, J. R., & Preston, K. F., (1991). Radical reactions of C60. *Science, 254*, 1183–1185.
48. Labhasetwar, V., Sog, C., Humphrey, W., Shebuski, R., & Levy, R. J., (1998). Arterial uptake of biodegradable nanoparticles: Effect of surface modifications. *Journal of Pharmaceutical Sciences, 87*, 1229–1234.
49. Lam, C. W., James, J. T., McCluskey, R., Arepalli, S., & Hunter, R. L., (2006). A review of carbon nanotube toxicity and assessment of potential occupational and environmental health risks. *Critical Reviews in Toxicology, 36*(3), 189–217.
50. Landsiedel, R., Kapp, M. D., Schulz, M., Wiench, K., & Oesch, F., (2009). Genotoxicity investigations on nanomaterials: Methods, preparation and characterization of test material, potential artifacts and limitations—many questions, some answers. *Mutat. Res., 681*, 241–258. Google Scholar Medline.
51. Liu, H., Liu, T., Wang, H., Li, L., Tan, L., Fu, C., Nie, G., Chen, D., & Tang, F., (2013). Impact of pegylation on the biological effects and light heat conversion efficiency of gold nanoshells on silica nanorattles. *Biomaterials, 34*, 6967–6975.
52. Lux, R., (2008). *The Nanotech Report* (4th edn., p. 246). Lux, New York.
53. MacNee, W., & Donaldson, K., (2000). How can ultrafine particles be responsible for increased mortality? *Monaldi. Arch. Chest. Dis., 55*, 135–139.
54. Magrez, A., Kasa, S., Salicio, V., Pasquier, N., Won, S. J., Celio, M., Catsicas, S., Schwaller, B., & Forro, L., (2006). Cellular toxicity of carbon-based nanomaterial. *Nano Letters, 6*(6), 1121–1125.
55. Majumder, D., Dutta, K. S., Goswami, A., & Banerjee, N., (2012). *Nanotoxicology: A Threat to the Environment and to Human Beings Proceedings of the International Symposium on Engineering Under Uncertainty: Safety Assessment and Management (ISEUSAM – 2012)*, pp. 385–400.
56. Maynard, A. D., (2006). *Nanotechnology: A Research Strategy for Addressing Risk.* PEN, Washington, D.C., Project on Emerging Nanotechnologies, Woodrow Wilson International Center for Scholars. http://web.pdx.edu/~pmoeck/phy381/nano%20 risks.pdf (Accessed on 5 November 2019).
57. Maynard, A. D., (2007). Nanotechnology: The next big thing, or much ado about nothing? *The Annals of Occupational Hygiene, 51*, 1–12. http://dx.doi.org/10.1093/ annhyg/mel071 (Accessed on 5 November 2019).
58. Maynard, A. D., (2009). New study seeks to link seven cases of occupational lung disease with nanoparticles and nanotechnology, 2020. *Science (Web Site).* http://2020science.org/2009/08/18/new-study-seeksto- link-seven-cases-of-ocupational-lung-disease-withnanoparticles- and-nanotechnology/ (Accessed on 5 November 2019).
59. Maynard, A. D., & Aitken, R. J., (2007). Assessing exposure to airborne nanomaterials: Current abilities and future requirements. *Nanotoxicology, 1*, 26–41. Google Scholar CrossRef.

60. Maynard, A. D., & Kuempel, E. D., (2005). Airborne nanostructured particles and occupational health. *J. Nanoparticle Res., 7*, 587–614. Google Scholar.
61. Mikawa, M., Kato, H., Okumura, M., Narazaki, M., Kanazawa, Y., Miwa, N., & Shinohara, H., (2001). *Bioconjug. Chem., 12*, 510.
62. Monteiro-Riviere, N. A., & Inman, A. O., (2006). Challenges for assessing carbon nanomaterial toxicity to the skin, *Carbon, 44*, 1070–1078.
63. Murashov, V., Engel, S., Savolainen, K., Fullam, B., Lee, M., & Kearns, P., (2009). Occupational safety and health in nanotechnology and Organization for Economic Cooperation and Development. *J. Nanopart. Res., 11,* 1587–1591. Google Scholar.
64. NANO—National Nanotechnology Initiative So What Is Nanotechnology? http://www.nano.gov/nanotech-101/what/definition (Accessed on 5 November 2019).
65. Nanotechnology and human health: Scientific evidence and risk governance Report of the WHO expert meeting, Bonn, Germany Publications WHO Regional Office for Europe UN City, Marmorvej 51 DK-2100. Copenhagen Ø, Denmark.
66. *Nanotechnology Web Page*, (2008). Department of Toxic Substances Control..
67. Nel, A., (2005). Air pollution-related illnesses: Effects of particles. *Science, 308*, 804–806. http://dx.doi.org/10.1126/science.1108752 (Accessed on 5 November 2019).
68. Nel, A., Xia, T., Madler, L., & Li, N., (2006). Toxic potential of materials at the nanolevel. *Science, 311*, 622–627.
69. Nohynek, G. J., Antignac, E., Re, T., & Toutain, H., (2010). Safety assessment of personal care products/cosmetics and their ingredients. *Toxicol. Appl. Pharmacol., 243*, 239–259. Google Scholar Medline.
70. Oberdorster, G., (2001). Pulmonary effects of inhaled ultrafine particles. *Int. Arch. Occup. Environ. Health, 74*, 1–8.
71. Oberdörster, E., (2004). Manufactured nanomaterials (Fullerenes, C60) induce oxidative stress in the brain of juvenile largemouth bass. *Environmental Health Perspectives, 112*(10), 1058–1062.
72. Oberdörster, G., Oberdörster, E., & Oberdörster, J., (2005). Nanotoxicology: An emerging discipline evolving from studies of ultrafine particles. *Environ Health Perspect., 113*, 823–839. Google Scholar Medline.
73. Oberdörster, G., Sharp, Z., Atudorei, V., Elder, A., Gelein, R., Kreyling, W., & Cox, C., (2004). Translocation of inhaled ultrafine particles to the brain. *Inhalation Toxicology, 16*, 437–445. Google Scholar Medline.
74. Oberdörster, G., Stone, V., & Donaldson, K., (2007). Toxicology of nanoparticles: A historical perspective. *Nanotoxicology, 1*, 2–25. Google Scholar.
75. Papp, T., Schiffmann, D., Weiss, D., Castranova, V., Vallyathan, V., & Rahman, Q., (2008). Human health implications of nanomaterial exposure. *Nanotoxicology, 2*(1), 9–27.
76. Peixel, T. S., Nascimento, E., De, S., Schofield, K. L., Arcuri, A. S. A., & Bulcão, R. P., (2015). *Nanotoxicology and Exposure in the Occupationa Setting Occupational Diseases and Environmental Medicine, 3*, 35–48.
77. Pekkanen, J., Peters, A., Hoek, G, Tiittanen, P., Brunekreff, B., De Hartog, J, Heinrich, J., Ibald-Mulli, A., Kreyling, W. G., Lanki, T., Timonen, K. L., & Vanninen, E., (2002). Particulate air pollution and risk of ST-segment depression during repeated submaximal exercise tests among subjects with coronary heart disease. The exposure

and risk assessment for fine and ultrafine particles in ambient air (ULTRA) study. *Circulation, 106*, 933–938.
78. Penttinen, P., Timonen, K. L., Tiittanen, P., Mirme, A., Ruuskanen, J., & Pekkanene, J., (2001). Ultrafine particles in urban air and respiratory health among adult asthmatics. *Eur. Resp. J., 17*(3), 428–435.
79. Perez, J. E., Contreras, M. F., Vilanova, E., Felix, L. P., Margineanu, M. B., Luongo, G., Porter, A. E., Dunlop, I. E., Ravasi, T., & Kosel, J., (2015). Cytotoxicity and intracellular dissolution of nickel nanowires. *Nanotoxicology, 22*, 1–38.
80. Pescovitz, D., (2006). *Magic Nano Recall.* Boing boing (web site). http://boingboing.net/2006/04/10/magicnano- recall.html (Accessed on 5 November 2019).
81. Peters, A., (2005). Particulate matter and heart disease: Evidence from epidemiological studies. *Toxicology and Applied Pharmacology, 207*, S477-S482.
82. Peters, A., Wichmann, H. E., Tuch, T., Heinrich, J., & Heyder, J., (1997). Respiratory effects are associated with the number of ultrafine particles. *Am. Resp. Crit. Care Med., 155*, 1376–1383.
83. Poland, C. A., Duffin, R., Kinloch, I., Maynard, A., Wallace, W. A. H., Seaton, A., Stone, V., Brown, S., MacNee, W., & Donaldson, K., (2008). Carbon nanotubes introduced into the abdominal cavity of mice show asbestos-like pathogenicity in a pilot study. *Nature Nanotechnology, 3*, 423–428. Published online: | doi:10.1038/nnano.2008.111.
84. Polshettiwar, V., Basset, J. M., & Astruc, D., (2012). Nanoscience makes catalysis greener. *Chem. Sus. Chem., 5*, 6–8.
85. Pope, C. A., Burnett, R. T., Thurston, G. D., Thun, M. J., Calle, E. E., Krewski, D., & Godleski, J. J., (2004). Cardiovascular mortality and long-term exposure to particulate air pollution: Epidemiological evidence of general pathophysiological pathways of disease. *Circulation, 109*, 71–77.
86. Qingnuan, L., Yan, X., Xiaodong, Z., Ruili, L., Quiqui, D., Xiaoguang, S., Shaoliang, C., & Wenxin, L., (2002). Preparation of ^{99m}Tc-C[60][OH][x] and its biodistribution studies. *Nucl. Med. Biol., 29*, 707.
87. The Royal Society and Royal Academy of Engineering, (2004). *Nanoscience and Nanotechnologies: Opportunities and Uncertainties*. London: The Royal Society. http://www.nanotec.org.uk/finalReport.htm (Accessed on 5 November 2019).
88. RSC—Royal Society of Chemistry, (2009). *The Future for Nanoscience and Nanotechnology.* http://www.iop.org/activity/policy/Publications/file_22332.pdf (Accessed on 5 November 2019).
89. Samet, J. M., Dominici, F., Curriero, F. C., Coursac, I., & Zeger, S. L., (2000). Fine particulate air pollution and mortality in 20 US cities, 1987–1994. *N. Engl. J. Med., 343*, 1742–1749.
90. SCENIHR—The Scientific Committee on Emerging and Newly Identified Health Risks, European Commission, (2006). *What is Nanotechnology?* http://ec.europa.eu/health/opinions2/en/nanotechnologies/l-3/1-introduction.htm (Accessed on 5 November 2019).
91. Schulte, P. A., & Buentello, F. S., (2007). Ethical and scientific issues of nanotechnology in the workplace. *Environmental Health Perspectives, 115*, 5–12.
92. Schulte, P. A., Iavicoli, I., Rantanen, J. H., Dahmann, D., Iavicoli, S., Pipke, R., Guseva, C. I., Boccuni, F., Ricci, M., Polci, M. L., Sabbioni, E., Pietroiusti, A., &

Mantovani, E., (2016). Assessing the protection of the nanomaterial workforce. *Nanotoxicology*. doi: 10.3109/17435390.2015.1132347.
93. Schwartz, J., (1994). Air pollution and daily mortality: A review and meta analysis. *Environ. Res., 64*, 36–52.
94. Schwartz, J., & Morris, R., (1995). Air pollution and hospital admissions for cardiovascular disease in Detroit, Michigan. *Am. J. Epidemiol., 142*, 23–35.
95. Seaton, A., MacNee, W., Donaldson, K., & Godden, D., (1995). Particulate air pollution and acute health effects. *Lancet, 345*, 176–178.
96. Sharma, M., (2010). Understanding the mechanism of toxicity of carbon nanoparticles in humans in the new millennium: Asystemic review. *Indian J. Occup. Environ. Med., 14*(1), 3–5.
97. Shimada, A., Kawamura, N., Okajima, M., Kaewamatawong, T., Inoue, H., & Morita, T., (2007). Translocation pathway of intratracheally instilled ultrafine particles from the lung into blood circulation in the mouse. *Toxicologic Pathology, 34*, 949–957.
98. Song, Y., Xue, L., & Du, X., (2009). Exposure to nanoparticles is related to pleural effusion, pulmonary fibrosis and granuloma. *European Respiratory J., 34*(3), 559–567.
99. Tinkle, S. S., Antonini, J. M., Rich, B. A., Roberts, J. R., Salmen, R., DePree, K., & Adkins, E. J., (2003). Skin as a route of exposure and sensitization in chronic beryllium disease. *Environ. Health Perspect., 111*, 1202–1208. Google Scholar Medline.
100. Tsuji, J. S., Maynard, A. D., Howard, P. C., James, J. T., Lam, C. W., Warheit, D. B., & Santamaria, A. B., (2006). Research strategies for safety evaluation of nanomaterials, part IV: Risk assessment of nanoparticles. *Toxicol. Sci., 89*, 42–50. Google Scholar Medline.
101. Utell, M. J., & Frampton, M. W., (2000). Acute health effects of ambient air pollution: The Ultrafine particle hypothesis. *J. Aerosol. Med., 13*, 355–359.
102. Viswanath, B., & Kim, S., (2016). Influence of nanotoxicity on human health and environment: The alternative strategies. Springer International Publishing Switzerland. In: De Voogt, P., (ed.), *Reviews of Environmental Contamination and Toxicology* (Vol. 242). doi: 10.1007/398_2016_12. From: https://link.springer.com/chapter/10.1007/398_2016_12.
103. Wang, H., Wang, J., Deng, X., Sun, H., Shi, Z., Gu, Z., Liu, Y., & Zhaoc, Y., (2004). Biodistribution of carbon single-wall carbon nanotubes in mice. *J. Nanosci. Nanotech., 4*(8), 1019–1024.
104. Wichmann, H. E., Spix, C., Tuch, T., Wölke, G., Peters, A., Heinrich, H., Kreyling, W. G., & Heyder, J., (2000). *Daily Mortality and Fine and Ultrafine Particles in Erfurt*. Germany. Part 1: Role of particle number and particle mass. HEI Research Report # 98: Health Effects Institute, Boston, Ma.
105. Wichmann, H. E., (2003a). What can we learn today from the central European smog episode of 1985 and earlier episodes? *Int. J. Hyg. Environ. Health, 206*, 505–520.
106. Wichmann, H. E., (2003b). Epidemiology of ultrafine particles. *BIA Report 7/2003e*, pp. 59–92.
107. Yuliang, Z., Bing, W., Weiyue, F., & Chunli, B. (2010–2011). *Nanotoxicology: Toxicological and Biological Activities of Nanomaterials* © Encyclopedia of Life Support Systems (EOLSS).

CHAPTER 9

Nanomaterials: Properties, Synthesis, Characterizations, and Toxicities

MOHIT AGARWAL,[1] RAJESH SINGH TOMAR,[1] R. K. SINGH,[2] ANIL KUMAR MEENA,[2] and ANURAG JYOTI[1,*]

[1]*Amity Institute of Biotechnology, Amity University Madhya Pradesh, Maharajpura, Gwalior – 474 005, India*

[2]*Division of Toxicology, CSIR-Central Drug Research Institute, Jankipuram Extension, Lucknow – 22 6031, India*

[*]*Corresponding author. E-mail: ajyoti@gwa.amity.edu*

ABSTRACT

This chapter represents a detailed overview of properties, synthesis, characterization, and toxicities related to nanoparticles (NPs). NPs are minute particles of 1–100 nm in size, and they are just a link between bulk materials and atomic or molecular structures. As compared to conventional bulk particles, nanomaterials exhibit different unique properties. They can be classified into different classes based on its properties, shapes, or sizes. NPs possess unique physical and chemical properties due to their high surface area and nanoscale size. Their optical properties are reported to be dependent on the size, which imparts different colors due to absorption in the visible region. Generally, NPs already exist in nature; natural sources of NPs in the atmosphere are volcanic eruptions, minerals such as clay and mica, natural colloids include blood and milk. The main approaches for NPs synthesis are either bottom-up or top-down approach, and various synthesis procedures include: physical, chemical, biological, and green synthesis. To understand the possible potential of NPs, a deeper knowledge of their synthesis and characterization is needed. Characterization is done by using a variety of different techniques. Nanoparticle characterization can be preliminarily done by UV-visible spectroscopy, and after preliminary confirmation, it can be done as needy, by using FTIR, TEM, XRD. Metallic NPs are reported to be so rigid and stable than their degradation is not easily achievable, which can lead to several toxicities. Toxicity can refer to the effect on a cell (cytotoxicity), gene (genotoxicity), an organ (e.g., renal or liver toxicity), or the whole organism and an estimate of how much of a substance causes a kind of harm.

9.1 INTRODUCTION: AN OVERVIEW

Nanotechnology is defined as any technology that utilizes nanomaterials, which are in the range from 1–100 nm. To know the world of nanotechnology, one should have knowledge about the units of measures involved. A nanometer (nm) is one-billionth of a meter and is still very large if we compare it to the atomic scale. For instance, cells are our nature's *nanomachines* [5]. Bulk materials possess continuous physical properties, and the same is applied to macro-sized material, but when they reach to nano-scale properties changes. Actually, these basically depend on the size and are different from the properties of bulk at macro scales. Nanomaterials have been found to have novel physical and chemical properties with respect to their large size counterparts. Also, the unique magnetic, electrical, optical, physicochemical properties of NPs arise due to its higher surface area to volume ratio.

9.2 UNIQUE CHARACTERISTICS/PROPERTIES OF NANOMATERIALS

Nanotechnology is a prominent area of science that is being explored in different field biotechnological, pharmacological, and in applied sciences. NPs are minute particles of 1–100 nm in size. They are just a link between bulk materials and atomic or molecular structures. NPs exhibit different properties when compared to conventional bulk particles. NPs have special optical properties as they can confine their electrons and also can perform better quantum effect due to the presence of surface plasmon resonance (SPR) [11, 14]. Color, NPs are of different colors, varying on metal salt utilized, on reducing and the capping agent being utilized. It has been reported that when gold materials are converted to nanomaterials, they turn into red color. Gold nanoparticles (AuNPs) interaction with light is vigorously governed by the particle sizes of the materials. The melting point also changes; it drastically falls down when the particle size reaches to nanoscale [24]. NPs possess mechanical vigor as compared to conventional counterparts. It is one or two times higher in magnitude than that of in bulk. Conversion of materials into nanoscale increases crystal perfection or reduction of defects, which would result in the enhancement of mechanical vigor. In general, the hardness of metals increases linearly with an increase of grain size. The electrical property also changes. By this, special property NPs can enhance crystal perfection. In integration, a reduction in particle size below a critical dimension, i.e., electron de Broglie wavelength would result in a modified electronic structure with a wide and discrete bandgap. It also shows good catalytic activity due to the immense surface, NPs composed of transition element exhibit intriguing catalytic properties. In special cases, catalysis may be enhanced and more concrete by embellishing these particles with gold or platinum clusters. In magnetic NPs, the energy may be that minute that the vector of magnetization fluctuates thermally called superparamagnetism. Physically contacting superparamagnetic particles is losing this special property by interaction, except the particles are kept at a distance. A consequential property of NPs is to compose suspensions. At elevated temperatures, especially, NPs possess the property of diffusion. NPs are generally based on their dimensionality. 1D nanomaterials, i.e., they are typically thin films or surface coatings, computer chips, and hard coatings on eyeglasses. These have been utilized in electronics, chemistry, and engineering. 2D nanomaterials include 2D nanostructured films, with nanostructures firmly affixed to a substrate, or nanopore filters. Asbestos fibers are an example of 2D NPs, and 3D nanomaterials, these include thin films deposited under atomic-scale porosity and colloids.

9.3 NANOPARTICLES (NPS)-BASED MATERIAL

9.3.1 NATURAL NANOMATERIALS

Natural nanomaterials means it belongs to the natural world without any anthropogenic modification or processing and should have a very unique property due to its inherent nanostructure. Natural sources of NPs in the atmosphere are volcanic eruptions [20]. There are many natural materials to which common peoples are familiar with their unique properties and composition. One of the best examples is the Lotus rolling activity for water droplets and dust removal. The nanostructure of this plant is responsible for its unique property and ability to self-clean. Exactly, for this reason, this plant is considered a *sign of purity*. Its leaves have a very outstanding property that it's totally repelling water because of its superhydrophobic nature. The basic mechanism behind is dragging the dirt away from the surface of the plant by rolling off water droplets. This mechanism is known as *the self-cleansing action of lotus*; due to this, the plant is resistant to dirt. This property was first investigated by Wilhelm Barthlott and published a paper describing the lotus effect (Figure 9.1).

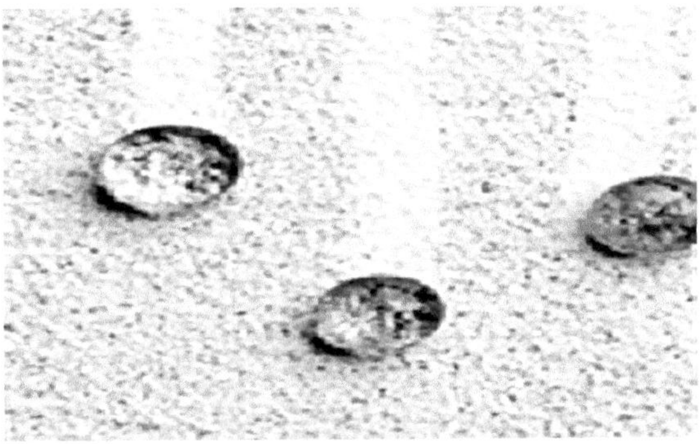

FIGURE 9.1 Assuming lotus flower with water droplets removing dirt by the rolling mechanism.

In nature, there are various outstanding phenomenon which can easily be shortlisted under natural nanomaterials phenomenon's which are:

1. NPs from natural erosion and volcanic activity.
2. Minerals such as clay and mica are the types of layered silicates that are characterized by a fine 2D crystal structure. Naturally occurring clay includes montmorillonite (MMT).
3. Natural colloids include blood, and milk is the best liquid colloid occurs naturally, fog, mist, and smog are aerosol type, while gelatin is gel type.

4. Materials like shells, corals, bones, skin, claws, feathers, horns, hair, etc. Some of these are made up of largely of very flexible protein, and some of these are formed with a polymer [6].

9.3.2 VARIOUS METALLIC NANOPARTICLES (NPS)

During the last decades, the biosynthesis of metallic NPs, such as zinc, silver, and AuNPs have received more attention.

9.3.2.1 ZINC OXIDE NANOPARTICLES (NPS)

Zinc oxide is listed as generally recognized as safe (GRAS) by U.S. Food and Drug Administration (FDA) (21CFR 182.8991). It is most commonly used as a food additive in the fortification of cereal-based food due to its antimicrobial property. It is also used in food can packing, meat, fish, and corn to preserve color. It is also famous for various applications such as photo-catalyst, optical material, cosmetics, UV absorber, and Gas sensor. While nano ranged of zinc oxide has more pronounced antibacterial activity than that of bulk, due to its large surface to volume ratio, which provides better binding with pathogens. Also, recent studies have shown selective toxicity to bacteria and minimal effect on human cells [16]. The bactericidal mechanism of ZnO NPs is complex and still under investigation [7].

9.3.2.2 SILVER NANOPARTICLES (AGNPs)

It is well known from ancient times that silver has a strong antimicrobial property [3]. To increase the antimicrobial effect of silver, it has to be converted into the nanoscale, as the size is reduced, surface area to volume ratio increases [1]. Silver nanoparticles (AgNPs) size of 10–100 nm has a strong bactericidal property [2, 4]. A broad range of nanosilver has been emerged. A more than thirteen hundred manufacture throughout the world are investing to develop nanotechnology-based products for the commercial market and more than 300 products made up of nanosilver. The world is growing well in nanotechnology, with an increase of 25% annually, about 3 trillion dollars by 2020 [18].

9.3.2.3 GOLD NANOPARTICLES (AUNPs)

Presently, the biosynthesis of gold nanoparticles (AuNPs) is an active research area. Syntheses of the gold nanoparticle by using biological as well as chemical approaches are well established. In this study, the biosynthesis of GNPs using lemongrass is well established. Although the mentioned plant, i.e., lemongrass study was earlier investigated, but it is by using AgNPs [9], but still there is no data available for the synthesis of GNPs by using lemongrass. The main objective of the present study is to develop a clean, non-toxic, and eco-friendly method for obtaining GNPs.

9.3.3 VARIOUS REDUCING AND CAPPING AGENT

Reducing agent is the chemical compound which performs reduction reaction. Typically reducing agents are used in the synthesis of metallic NPs. They generally reduce metal salts into pure metals. The amount of reducing agent required in a typical reaction is normally decided by the quantity of metal salt at the starting point. For example, sodium borohydride, amino acids, CTAB, ascorbic acid, etc., are reducing agents. While tri-sodium Citrate is reducing as well as a capping agent to prevent the binding. A capping agent is used for the stabilization of NPs is strongly absorbed monolayer of organic molecules. Particles can be functionalized using capping agents to impart useful properties. A capping agent can be polymerized to form a functional polymer and used to protect the surface of materials. Also, the capping agents like dendrimers, even clay particles protect the NPs from aggregation, and chemical reaction, for example, hydrobenzamide, citrate, polyvinyl pyrrolidone (PVP), mercaptoethanol, and thiourea is also a capping agent (Table 9.1).

TABLE 9.1 Precursors, Reducing Agent Polymeric Stabilizer's Used for Nanoparticles Synthesis

Sr. No.	Category	Salt Used
1	Precursors	Metal salt: Zinc acetate, copper sulfate, silver nitrate, auric chloride, etc.
2	Reducing Agent	Sodium citrate, citric acid, ascorbic acid, glucose, plants extract, etc.
3	(Polymeric) Stabilizer	PVP, PVA, plants extract, glucose etc.

9.4 DIFFERENT APPROACHES AND SYNTHESIS PROCEDURE

The main approaches for NPs synthesis are either bottom-up or top-down approach. The various synthesis procedures include: Physical, chemical, biological, and green synthesis. Broadly they can be classified hereunder.

9.4.1 TOP-DOWN APPROACHES AND BOTTOM-UP APPROACHES

The top-down approach involves the breaking down of bulk to nano-sized structure, and the bottom-up approach is to build material by atom by atom or by molecules by molecules (Figure 9.2).

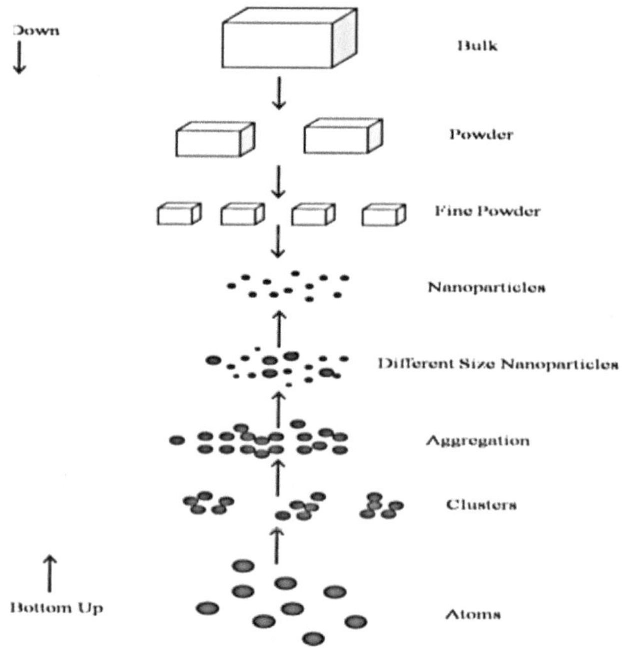

FIGURE 9.2 Top-down and bottom-up approaches.

9.4.2 VARIOUS SYNTHESIS PROCEDURES

Various techniques have been used for the synthesis of NPs. Broadly they may be classified under three main categories, i.e., physical, chemical, biological, and green synthesis.

9.4.2.1 PHYSICAL METHODS

There are different methods for the synthesis but using physical methods. The two are well-known evaporation-condensation and laser ablation methods [21, 12].

9.4.2.2 CHEMICAL METHODS

The common approach for the synthesis of NPs is chemical reduction by using organic or inorganic reducing agents. Various different reducing agents are used for the synthesis, such as sodium citrate, ascorbic acid, sodium boro-hydride ($NaBH_4$), etc. is used for the reduction of bulk to ionic form in aqueous or non-aqueous solutions (Figure 9.3) [10].

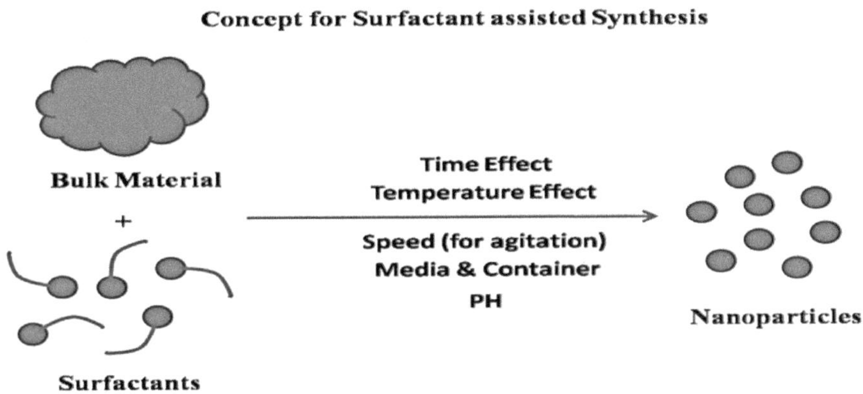

FIGURE 9.3 Concept for chemical nanoparticle synthesis.

9.4.2.3 BIOLOGICAL

Chemical approaches are not eco-friendly, while using these approaches might be quite helpful for the environment, while using biological agents for synthesis such as bacteria, fungi, algae for the synthesis are nontoxic to us and are cost-effective also. Bacteria have been used for the synthesis of NPs, and various reports suggest that highly stable silver NPs could be synthesized using the bioreduction of aqueous silver ions with culture supernatant of the non-pathogenic bacterium, *Bacillus licheniformis* [13], and *Fusarium oxysporum* [8].

9.4.2.4 GREEN SYNTHESIS

Green syntheses of NPs using plant parts such as leaves, stem, flower, and roots are very cost-effective, and less toxic. Due to easy availability plants, this method could be used for large-scale synthesis of highly stable mono-dispersed NPs [23]. As few papers were reported for the synthesis of AgNPs, *Camellia sinensis* (green tea) extract has been used [26]. NPs (5–50 nm) could be synthesized extracellularly using fungi, such as *Fusarium oxysporum*, with no evidence of flocculation of the particles even a month after the reaction [8]. The long-term stability of the nanoparticle solution might be due to the stabilization of the particles by proteins. A few reports are available regarding gold accumulation using algal genera, including cyanobacteria as a biological reagent. Cyanobacteria and eukaryotic alga genera such as *Lyngbya majuscule*, *Spirulina subsalsa*, *Rhizoclonium heiroglyphicum*, *Chlorella vulgaris*, and *Cladophora prolifera* can be used for the formation of gold NPs [17]. Synthesis of NPs using plants is very cost-effective and thus can be used as an economical and valuable alternative for the large-scale production of NPs. *Camellia sinensis* (green tea) extract has been used as a reducing and stabilizing agent for the biosynthesis of silver NPs [26].

9.5 NANOMATERIALS CHARACTERIZATION

To understand the possible potential of NPs, a deeper knowledge of their synthesis and characterization is needed. Characterization is done by using a variety of different techniques. Nanoparticle characterization can be preliminarily done by UV-visible spectroscopy, and after preliminary confirmation, it can be done as needy by using FTIR, TEM, XRD, etc.

9.5.1 UV-VISIBLE SPECTROSCOPY

UV-visible spectroscopy is a most simplified technique, is used to quantify the amount of light scattered by the sample. Briefly, a sample is placed between a light source and a photo-detector, and the intensity of a beam of light is measured before and after passing through the sample. [15]. NPs have optical properties that are sensitive to size, shape, concentration, agglomeration state, and refractive index near the nanoparticle surface, which makes UV-visible spectroscopy a valuable tool for identifying, characterize them.

9.5.2 ELECTRON MICROSCOPY

Electron microscopy is one of the best tools for identifying the material in nanoscale. It basically focuses on a beam of electrons that accelerate from an electron gun. Generally, an electron microscope can resolve up to 2A°. Transmission electron microscopy (TEM) and scanning electron microscopy (SEM) are considered the gold standard for nanoparticle characterization [22, 19]. SEM and TEM both offer unique benefits for nanoparticle characterization, and it offers researchers a very detailed images of specimens at microscopic and nanoscale.

9.5.3 XRD

X-ray diffraction patterns have been widely used in nanoparticle as a primary characterization tool to determine the crystal structure and crystallite size. The simplest and most widely used method for estimating the average crystallite size is from the full width at half maximum (FWHM) of a diffraction peak.

9.6 TOXICITY

Toxicity is the degree to which a substance (a toxin or poison) can harm humans or animals. Toxicity can refer to the effect on a cell (cytotoxicity), gene (genotoxicity), an organ (e.g., renal or liver toxicity), or the whole organism and an estimate of how much of a substance causes a kind of harm. Basically, all substances are potentially toxic depending on the quantity, such as oxygen or water being the simplest examples.

9.6.1 CYTOTOXICITY ASSAY

Basically, to access the toxicity profile on a cellular system *in-vitro*, we need to use cell culture assays. Cytotoxicity assays are the basic potential link between the *in-vitro* and *in-vivo* system so as to move further with the compound or to stop. These include colorimetric assay (MTT), enzymatic activity released by viable cells (LDH), SRB assay, bioluminescent methods (luciferase enzyme) (Neutral red and Trypan blue), and several others.

9.6.2 GENOTOXICITY ASSAY

The chromosomal aberrations in Swiss mice bone marrow cells for the exposure of synthesized NPs, whose genotoxic potential (if any) has to be established as per regulatory requirement (OECD Guidelines). A chromosome shows aberration, which may be structural or numerical. Detection of aberration in the chromosome, therefore, indicates the Genotoxicity of a chemical. Chromosomal aberrations are generally evaluated in first post-treatment mitosis following the administration of test items to laboratory mice.

The micronucleus assay is a mutagenic test system for the detection of chemicals that induce the formation of small membrane-bound DNA fragments, i.e., micronuclei in the cytoplasm of the interphase cells. These micronuclei may originate from acentric fragments or whole chromosomes that are unable to migrate with the rest of the chromosome during the anaphase of cell division.

The purpose of the micronucleus assay is to detect those agents which modify chromosomes structure, and segregation is such a way as to lead to the induction of micronuclei in interphase cells.

The study of micronucleus study in Swiss mice bone marrow cells following exposure of test item, whose genotoxicity potential (if any) has to be established as per regulatory (OECD Guidelines for testing chemicals).

9.6.3 ACUTE AND SUB-ACUTE TOXICITY

The globally harmonized system (GHS) defines acute toxicity as "those adverse effects occurring following oral administration of a single dose of a substance, or multiple doses given within 24 hours, or an inhalation exposure of 4 hours." It is the first test conducted and provides critical information on the relative toxicity likely to arise from a single exposure.

The Organization for Economic Cooperation and Development (OECD) elaborate five test guidelines for describing acute systemic testing.

Sub-acute toxicity test determines toxicity from exposure for a substantial portion of a subject's life, and the GHS defines it as "specific target organ/systemic toxicity arising from repeated exposure." In rats, these studies range in duration from 28-days (sub-acute studies) to 90-days (sub-chronic studies), and even 12-months (chronic studies), and consist of repeated doses in oral, inhalation, and dermal administration.

KEYWORDS

- **generally recognized as safe**
- **gold nanoparticles**
- **montmorillonite**
- **polyvinyl pyrrolidone**
- **scanning electron microscopy**
- **transmission electron microscopy**

REFERENCES

1. Agnihotri, S., Mukherji, S., & Mukherji, S., (2014). Size-controlled silver nanoparticles synthesized over the range 5–100 nm using the same protocol and their antibacterial efficacy, *RSC Adv.*, *4*, 3974–3983.
2. Chandrakanth, R. K., Ashajyothi, C., Oli, A. K., & Prabhurajeshwar, C., (2014). Potential bactericidal effect of silver nanoparticles synthesized from Enterococcus species. *Orient J. Chem.*, *30*(**3**).
3. Chen, X., & Schluesener, H. J., (2008). Nanosilver: A nanoproduct in medical application. *Toxicology Letters*, *176*, 1–12.
4. Daniel, J. F. *PEN 19 – Voluntary Initiatives, Regulation, and Nanotechnology Oversight*. http://www.nanotechproject.org (Accessed on 5 November 2019).
5. Filipponi, L., & Sutherland, D., (2010). *Nanoscience in Nature, Fundamental Concepts of Nanoscience and Nanotechnology*. Aarhus University, Denmark.
6. Fillipponi, L., & Sutherland, D., (2010). Introduction to nanoscience and nanotechnology. *Fundamental Concepts of Nanoscience and Nanotechnology*, Aarhus University, Denmark.
7. Huang, Z., Zheng, X., Yan, D., Yin, G., Liao, X., Kang, Y., Yao, Y., Huang, D., & Hao, B., (2008). Toxicological effect of ZnO nanoparticles based on bacteria. *Langmuir*, *24*(8), 4140–4144.

8. Iravani, S., Korbekandi, H., Mirmohammadi, S. V., & Zolfaghari, B., (2014). Synthesis of silver nanoparticles: Chemical, physical and biological methods. *Res. Pharm. Sci., 9*(6), 385–406.
9. Kruis, F., Fissan, H., & Rellinghaus, B., (2000). Sintering and evaporation characteristics of gas-phase synthesis of size-selected PbS nanoparticles. *Mater. Sci. Eng. B., 69*, 329–334.
10. Lengke, M., Ravel, B., Fleet, M. E., Wanger, G., Gordon, R. A., & Southam, G., (2006). Mechanisms of gold bioaccumulation by filamentous cyanobacteria from gold (III)-chloride complex. *Environ. Sci. Technol., 40*, 6304–6309.
11. Lubick, N., & Betts, K., (2008). Silver socks have cloudy lining. *Environ. Sci. Technol., 42*(11).
12. Merga, G., Wilson, R., Lynn, G., Milosavljevic, B., & Meisel, D., (2007). Redox catalysis on "naked" silver nanoparticles. *J. Phys. Chem. C., 111*, 12220–12206.
13. Niu, H., & Volesky, B., (2000). Gold-cyanide biosorption with L-cysteine. *J. Chem. Technol. Biotechnol., 75*, 436–442.
14. Kumbhakar, P., Ray, S. S., & Stepanov, A. L., (2014). Optical properties of nanoparticles and nanocomposites, *Journal of Nanomaterials,* 2 Article ID 181365. http://dx.doi.org/10.1155/2014/181365 (Accessed on 5 November 2019).
15. Rades, S., Hodoroaba, V. D., Salge, T., Wirth, T., Lobera, M. P., Labrador, R. H., Natte, K., Behnke, T., Gross, T., & Unger, W. E. S., (2014). High-resolution imaging with SEM/T-SEM, EDX and SAM as a combined methodical approach for morphological and elemental analyses of single engineered nanoparticles. *RSC Adv., 4*, 49577–49587.
16. Reddy, K. M., Feris, K., Bell, J., Wingett, D. G, Hanley, C., & Punnoose, A., (2007). Selective toxicity of zinc oxide nanoparticles to prokaryotic and eukaryotic systems, *Appl. Phys. Lett., 24*, 90.
17. Ronson, C. T., Ste, K., & San, D., (2012). *Nano Composix: UV/Vis/IR Spectroscopy Analysis of Nanoparticles, 1*(1), 1–6.
18. Shalaka, A. M., Pratik, R. C., Vrishali, B. S., & Suresh, P. K., (2011). Rapid biosynthesis of silver nanoparticles using cymbopogan citratus (lemongrass) and its antimicrobial activity. *Nano-Micro Letters, 3*(3), pp. 189–194. Publication Date (Web): (Article). doi: 10.3786/nml.v3i3.p189–194.
19. Silvestre, C., & Cimmino, S., (2013). 'Ecosustainable polymer nanomaterials for food packaging: Innovative solutions, characterization needs. *Safety and Environmental Issues* (p. 235). CRC Press.
20. Strambeanu, N., Demetrovici, L., & Dragos, D., (2015). *Natural Sources of Nanoparticles*. Springer International Publishing, Switzerland.
21. Tsuji, T., Iryo, K., Watanabe, N., & Tsuji, M., (2002). Preparation of silver nanoparticles by laser ablation in solution: Influence of laser wavelength on particle size. *Appl. Surf. Sci., 202*, 80–85.
22. Yu, L. L., & Wood, L. J. S. E., (2010). Long. NIST – NCL joint assay protocol, PCC-15 version 1.1. *Measuring the Size of Colloidal Gold Nano-Particles Using High-Resolution Scanning Electron Microscopy, 1*(1), 1–21.
23. US FDA Guidance for Industry Analytical Procedures and Methods Validation Chemistry, (2015). *Manufacturing, and Controls Documentation*, 1–15.

24. Vilchis-Nestor, A. R., Sánchez-Mendieta, V., Camacho-López, M. A., Gómez-Espinosa, R. M., Camacho-López, M. A., & Arenas-Alatorre, J., (2008). Solventless synthesis and optical properties of Au and Ag nanoparticles using *Camellia sinensis* extract. *Materials Letters, 62*, 3103–3105.
25. Vollath, D., (2008). *Nanomaterials: An Introduction to Synthesis, Properties and Application*, pp. 865–870.
26. Begum, N.A., Mondal, S., Basu, S., Laskar, R.A. & Mandal, D. (2009). Biogenic synthesis of Au and Ag nanoparticles using aqueous solutions of Black Tea leaf extracts. *Colloids Surf B Biointerfaces., 71,* 113–118.

CHAPTER 10

Nanoparticles: Method of Production and Future Prospective

NEHA SHARMA[1,*] and D. S. RATHORE[2]

[1]*Amity Institute of Biotechnology, Amity University Madhya Pradesh, Gwalior, India*

[2]*Department of Biotechnology, Government Kamla Raja Girls P. G. (Autonomous) College, Gwalior, Madhya Pradesh, India*

**Corresponding author. E-mail: drneha16may@gmail.com*

ABSTRACT

Nanotechnology techniques have become popular day-by-day. Many physical and chemical compounds are used in synthesis of nanoparticles with different solvent system, but the use of such methods are harmful to healthy cells due to toxicity of byproduct, so various plant extract could be used for the synthesis of nanoparticles. Extract of plants have the property to act as both reducing agents and stabilizing agents in the synthesis of NPs with metals. Metals like silver and gold are one of the most attractive and useful for synthesis of nanoparticles because it have antimicrobial efficiency and biomedical properties. Synthesis of silver nanoparticles (AuNPs) with plant extract has an advantage over the other physical methods as it is safe, green synthesis, eco-friendly, and simple to use.

10.1 INTRODUCTION

Nanotechnology has become one of the most popular technologies applied in all areas of science. Metal nanoparticles (NPs) produced by nanotechnology have received global attention due to their extensive applications in the biomedical and physiochemical fields. Plant extract and their effect on microorganisms have been extensively studied for synthesis of metal NPs and recognized as a green and efficient way for further exploiting antimicrobial agents as convenient nanofactories.

The term first introduced by Prof. Norio Taniguchi in 1974. He introduces the multidisciplinary nanotechnology covering the field of research and technology from physics, chemistry, and biology. The synthesis of NPs has introduced nanotechnology during the last two decades that produced novel compounds applied in various field of diagnostics, antimicrobial agents, drug delivery, textiles (clothing), electronics, bio-sensing, food industry, paints, cosmetics, medical devices and treatment of several chronic and acute disease such as cancer and AIDS.

Silver, gold, platinum, palladium, copper, zinc, and iron are the noble metals used in the synthesis of NPs of nano size. The application of NPs depends on properties such as size, shape, composition, crystalline nature, and structure. The NPs are metal atom clusters of range 1–100 nm, highly promising due to their wide range of applications in commercial products. The metal NPs are synthesized by various methods like physical, chemical, and biological. The biological synthesis of NPs involves Algae, actinomycetes, bacteria, fungi, and plants (Table 10.1) [9].

Table 10.1 Different Types of Metal Nanoparticles and Their Fields of Applications

S. No.	Metal Nanoparticle	Applications
1	Aluminum, Cesium, Cobalt, Copper	Industries: (Electronics, food & feed, space, chemical, textile, fertilizers, and pesticides)
2	Gold, Magnetite	Agriculture, biomedical, cosmetics, diagnostics, drug delivery
3	Nickel, Palladium	Energy science, environment, healthcare, light emitters, mechanics
4	Platinum, Silicon	Medical devices, nonlinear optical devices, paints, and coatings
5	Silver, Zinc	Pharmacological, photo-electrochemical, phytomining phytoremediation, sensors, and tracers, single-electron transistors, wastewater treatment

Increasing microbial resistance against the antibiotics and remarkable antimicrobial effect of metallic NPs on this resistance strain is an interesting area for researchers [2].

For ancient times, plants have been a valuable source of natural products for maintaining human health, with more intensive studies for natural therapies. Today, the use of phytochemicals for the pharmaceutical purpose has gradually increased in many countries. According to WHO, medicinal plants would be the best source to obtain a variety of drugs. In developed countries, about 80% of individuals use traditional medicine, which has phytocompounds derived from different parts of medicinal plants [7].

The phytochemical present in a crude extract of various parts of plants with known antimicrobial properties has great significance in the therapeutic treatments. In recent years, a number of studies have been conducted in various countries to prove the efficiency of photochemical. Many plants have been produced many secondary metabolites, which are used as an antimicrobial agent. These products are known by their active substances like, phenolic compounds which are part of the essential oils, as well as in tanning. Due to the broad resistance of microorganism against antibiotics used for human pathogens in recent years motivated to scientist to search for new antimicrobial substances from various sources like the medicinal plants [11]. It is well defined that the plant kingdom is an inexhaustible source of active phytoingredients valuable in the management of much intractable pathogenic disease [15].

Secondary metabolites as bioactive compounds are accumulated in all cells of the plant, but their concentration varies according to the plant parts, season climate, and particular growth phase. The leaf is one of the good sources of such compounds, and people are generally preferred for therapeutic purposes. Some of the active compounds inhibit the growth of disease-causing microbes either single or in combination [6].

Nature has long been an important source of medicinal agents. Natural sources have been used to synthesize a number of effective drugs, based on their use in traditional medicine. The plants have been used traditionally for the synthesis of pharmacologically active components from many centuries, and modern scientific studies have shown the existence of a good correlation between the traditional or folkloric application of some of the plants with antimicrobial activity [8].

10.2 METHODS FOR SYNTHESIS OF NANOPARTICLES (NPS)

Many physical and chemical methods used to synthesis of NPs with different solvent, but the use of such methods are harmful due to toxicity of byproduct in one or the many ways, so plants extract could be used for the synthesis of NPs.

10.3 SYNTHESIS OF NANOPARTICLES (NPS) USING PLANT EXTRACTS

The synthesis of phyto NPs is an emerging field of nanotechnology and biotechnology. Due to a growing need to develop environmental-friendly technologies in material synthesis, it has received increased attention [13]. Antimicrobial activity of phytocompound has motivated the researchers to synthesis the NPs using plant extract that allows better control of pathogenic microorganisms.

10.3.1 SYNTHESIS OF SILVER NANOMATERIAL

Nanotechnology is a part of biotechnology and an emerging field in the area of interdisciplinary research. Many chemical and physical methods are used in the synthesis of silver nanomaterials (AgNPs), but the development of reliable technology to produce NPs from Ag is an important phase of nanotechnology [10]. Several chemical methods have been developed for the synthesis of AgNPs. It includes a reduction in aqueous and nonaqueous solution, ultrasonic-assisted reduction, photo-induced or photo-catalytic reduction, microwave-assisted synthesis, irradiation reduction, the microemulsion method, etc. But chemical methods have various drawbacks, because of the use of various toxic solvents as well as the production of hazardous toxic byproducts, and use of high energy, which cause potential risks to human health and also to the environment [3, 5, 14]. Presently, there is an emergent need to develop an environment-friendly nanoparticle that does not use toxic chemicals in the process of synthesis. For synthesis, microbial-mediated biological metallic NPs have recently been recognized as a hopeful source for NPs [4].

The extraction of various plants was prepared by soxhlet method using different chemicals like ethanol, acetone, etc., as well as water and stores it for the synthesis of AgNPs. Aqueous solution of 10,000 ppm mol/L of AgNO3 was prepared. The extraction was added gradually to different flasks containing $AgNO_3$ for bioreduction. The volume ratio of extraction to aqueous $AgNO_3$ was 1:10. After some time, the plant extraction to aqueous $AgNO_3$ resulted in a change of color within the formation of AgNPs, showing its signatory color. The bioreduction of Ag+ ions was monitored by periodic sampling by the UV spectrophotometry.

10.3.1.1 CHARACTERIZATION OF THE SYNTHESIZED SILVER NANOPARTICLES (AGNPs)

UV-vis spectrophotometer were used to recorded optical absorbance (Systronics 2202 double beam model) in 200–800 nm wavelength range. It was observed that upon addition of the extract into the flask containing the aqueous silver nitrate solution, the color of the medium change the brown within two minutes. Change in the color indicates the formation of silver nanoparticles [1].

10.3.2 SYNTHESIS OF GOLD NANOPARTICLES (AUNPs) USING PLANT EXTRACTS

Biosynthesis of gold nanoparticles (AuNPs) using extraction of plant parts is getting more popular due to the strong antibacterial action against pathogenic microorganism. As it is well-known, gold has extremely reducing power; thus the synthesis of AuNPs using plant extract is useful not only because of its reducing property, but also because it can produce large quantities of NPs. Extraction of plant parts may act both as reducing agents and stabilizing agents in the synthesis of NPs. Extraction from various plant parts such as leaf, stem, and fruit was prepared through soxhlet method, and Gold ion solution was prepared by diluting 16.6 μL of 30 wt% $HAuCl_4$ solution (Sigma-Aldrich, St Louis, MO, USA) in 50 mL of DDW to form a 0.1 g/L solution. GNPs were prepared by mixing 10 mL of the gold ion solution (1.0 mg $HAuCl_4$) with 0.75 mL of plant extract. In the case of GNP, the mixture changed color to a deep purple/red within a few seconds. The color change was due to GNP formation, as verified by UV-vis absorption at 530 nm [12].

10.4 ADVANTAGES OF NANOTECHNOLOGY

As any other technique nanotechnology has many advantages and disadvantages, the advantages with this technology are as follows:

- Nanotechnology is widely applicable to developed electronic products include nano transistors, nano diodes, OLED, plasma displays, quantum computers, and many more.
- Energy sector is also benefited by nanotechnology. The development of more effective energy-producing, energy-absorbing, and energy storage products in smaller and more efficient devices are possible due to nanotechnology. Synthesis of some small items like batteries, fuel cells, and solar cells can be built by different techniques but can be made to be more effective with this technology.
- Nanotechnology has another industrial benefit is the manufacturing sector that will need materials like nanotubes, aerogels, nanoparticles, and other similar items. Nanotechnology makes these materials stronger, more durable, and lighter than those that are not produced with the help of nanotechnology.
- In the medical world, nanotechnology is also seen as a boon since these can help with creating nanodrugs, is also called smart drugs. These help to cure people faster and without the side effects that other traditional drugs have. We will also find that the research of nanotechnology in medicine is now focusing on areas like tissue regeneration, bone repair, immunity, and even cures for such ailments like cancer, diabetes, and other life threatening diseases.

10.5 DISADVANTAGES OF NANOTECHNOLOGY

We will also need to point out the disadvantages or negative side of nanotechnology:

- Due to nanotechnology, it is the possible loss of jobs in the traditional farming and manufacturing industry.
- It is more accessible to make atomic weapons and made to be more powerful and more destructive just because of nanotechnology.
- Since these particles are very small, problems can actually arise from the inhalation of these minute particles, much like the problems a person gets from inhaling minute asbestos particles.
- Production from nanotechnology is very expensive. It is also a little difficult to manufacture the products with nanotechnology.

10.6 FUTURE PROSPECTIVE

The development of multidrug resistant microbial strains and the appearance of strains which reduced susceptibility to antibiotics are continuously increasing. Such an increase has been attributed to indiscriminate use of broad-spectrum antibiotics, immunosuppressive agents, intravenous catheters organ transplantation, and ongoing epidermis of human immunodeficiency virus (HIV) infections. The problem of multidrug resistance promotes to search for new antimicrobial substances from various sources like medicinal plants.

The plants have usually provided a source of novel drug or phytocompounds, as plant herbal mixtures have made large contributions to human health. The use of plant extracts with known antimicrobial properties can be of great significance for therapeutic treatment [11]. Considering the vast potentiality of plants as sources for antimicrobial drugs with reference to antibacterial agents, a systematic investigation was undertaken to screen the plants for antibacterial activity and production of NPs.

Extraction of various part of plant has the property to act as both reducing agents and stabilizing agents in the synthesis of NPs with metals. Metals like silver and gold are one of the most attractive and useful NPs because of various properties like antimicrobial efficiency and biomedical properties. Synthesis of AuNPs with plant extract has an advantage over the other physical methods as it is safe, eco-friendly, and simple to use. Plants extracts have huge potential for the production of AuNPs of wide potential of applications with desired shape and size. A detailed study is required to give a proper mechanism of AuNPs biosynthesis using phytomolecules present in different plant extracts, which will be helpful to improve the properties of AuNPs.

10.7 CONCLUSION

For many years ago, people have used silver for its antibacterial qualities due to different properties. In the case of silver and gold phytonanoparticles, the antibacterial effect is greatly enhanced by nanotechnology because of their tiny size. NPs usually have better or different qualities than the bulk material of the same element. NPs have an immense surface area relative to volume. Therefore, NPs synthesized by silver and gold can increase the antimicrobial effects of its host material (plant extract) due to nanotechnology.

NPs are the interesting field and requirement of the development of nanomedicine and bionanotechnology. Silver and AuNPs produced from different parts of the plant have received considerable attention owing to their attractive phytochemical and physicochemical properties. Phyto NPs worked against human pathogens more efficiently and utilized as therapeutic tools. It is essential to understand the properties of NPs and their effect on microbes to evaluate the clinical application.

Ag and Au NPs produced from various parts of plant extract have already been tested in various pathogenic microorganisms like Gram (+) and Gram (–) bacteria as an antimicrobial compound by researchers. Today, one of the emerging problems with a large number of the bacteria has yet developed resistance to antibiotics. In the future, increasing bacterial resistance is a major problem and needs to develop a substitute for antibiotics. Phyto-NPs are an attractive alternative because they are nontoxic to the human body at low concentrations and having broad-spectrum antibacterial nature. Phyto-SNP worked at very low concentrations and significantly inhibits the growth of pathogenic microorganisms compared to antibiotics with no side effects.

It is concluded in this chapter that it would lead to the establishment of some valuable phyto nanocompound that has to be used to formulate new, different, and more potential antimicrobial drugs of natural origin with no side effects. More studies are required to identify the biologically active phytocompounds and to evaluate the efficiency of the compound against pathogenic microorganisms associated with various human diseases.

KEYWORDS

- **antimicrobial compound**
- **gold nanoparticles**
- **human immunodeficiency virus**
- **nanoparticles**
- **phytocompound**
- **silver nanoparticles**

REFERENCES

1. Ahmad, N., & Sharma, S., (2012). Green synthesis of silver nanoparticles using extracts of *Ananas comosus*. *Green and Sustainable Chemistry, 2*, 141–147.
2. Ahmed, S., Ahmad, M., Swami, B. L., & Ikram, S., (2015). A review on plants extract mediated synthesis of silver nanoparticles for antimicrobial applications: A green expertise. *Journal of Advanced Research, 7*, 17–28.
3. Awwad, A. M., Salem, N. M., & Abdeen, A. O., (2013). Green synthesis of silver nanoparticles using carob leaf extract and its antibacterial activity. *Int. J. Indus Chem., 4*, 1–6.
4. Badr, Y., Wahed, E., & Mahmoud, M. G., (2008). Photo catalytic degradation of methyl red dye by silica nanoparticles. *J. Haz. Mat., 154*, 245–253.
5. Bar, H., Bhui, D. K., Sahoo, G. P., Sarkar, P., Priyanka, S., et al., (2009). Green synthesis of silver nanoparticles using seed extract of *Jatropha curcas*. *Colloids Surf A Physicochem. Eng. Asp., 348*, 212–216.
6. Dhia, H., & Abeer, K., (2006). *J. Bio. Sci., 6*(1), 109–114.
7. Ellof. J. N., (1998). *J. Ethanopharmacol., 60*, 1–6.
8. Egharevba, H. O., & Kunle, O. F., (2010). Preliminary phytochemical and proximate analysis of the leaves of *Piliostigma thioniningii* (schumach) Mile Redhead. *Ethanobotanical Leaflets, 14*, 570–577.
9. Haleemkhan, A. A., Naseem, V., & Vardhini, B., (2015). Synthesis of nanoparticals from plants extract. *International Journal of Modern Chemistry and Applied Science, 2*(3), 195–203. ISSN 2349–0594.
10. Huang, Z., Jiang, X., Guo, D., & Gu, N., (2011). Controllable synthesis and biomedical applications of silver nanomaterials. *J. Nanosci. Nanotechnol., 11*, 9395–9408.
11. Iwu, M. W., Duncan, A. R., & Okunji, C. O. (1999). New antimicrobials of plant origin In: Janick, J. (Ed.), *Perspectives on New Crops and New Uses*. ASHS Press, Alexandria, VA, pp. 457–462.
12. Paz, E., Raya, Z., Sharon, H., Sofiya, K., Zeev, P., & Yehuda, Z., (2014). Green synthesis of gold nanoparticles using plant extracts as reducing agents. *International Journal of Nanomedicine, 9*, 4007–4021.
13. Sahayaraj, K., & Rajesh, S., (2011). Bionanoparticles: Synthesis and antimicrobial applications. *Sci. Against Microb. Pathog.*, 228–244.
14. Sathyavathi, R., Krishna, M. B., Rao, S. V., Saritha, R., & Rao, D. N., (2010). Biosynthesis of silver nanoparticles using *Coriandrum sativum* leaf extract and their application in nonlinear optics. *Adv. Sci. Lett., 3*, 138–143.
15. Shariff, Z. U., (2001). Feeling Nature's PAINS: Natural Products, Natural Product Drugs, and Pan Assay Interference Compounds (PAINS). *J. Nat. Products, 1*, 79–84.

CHAPTER 11

Nanoinformatics: An Emerging Trend in Cancer Therapeutics

MEDHA PANDYA,[1] SNEHAL JANI,[2] VISHAKHA DAVE,[3] and RAKESH RAWAL[4]

[1]*Assistant Professor, The K.P.E.S. Science Collage, M.K. Bhavnagar University, Bhavnagar, India, Tel.: +91-9662031001, E-mail: megsp85@gmail.com*

[2]*Assistant Professor, Amity School of Applied Sciences, Amity University Madhya Pradesh, Gwalior, India, Tel.: +91-8866014107, E-mails: sneh.jani@gmail.com; scjani@gwa.amity.edu*

[3]*Research Scholar, Department of Physics, M.K. Bhavnagar University, Bhavnagar, India, Tel.: +91-9408728323, E-mail: Vishakhadave1612@gmail.com*

[4]*Professor, Department of Life Sciences, Gujarat University, Ahmadabad, India, Tel.: +91-9925244855, E-mail: rakeshmrawal@gmail.com*

ABSTRACT

The current research in cancer nanotechnology focused on understanding the interactions of materials with the biological system at the molecular level. The technique of inventing any novel material to treat any carcinogenic condition requires the focus of genetic variations and understanding of data analysis. For example, the change in gene at the nucleotide level has a notable influence on the protein to transform in the fusion protein. Many computational methods can predict the structures of fusion genes and proteins that contribute to cancer progression. The current study focused on identifying fusion protein and developing related targeted molecules to design suitable therapeutics in particular malignancy. The nanotechnology-based delivery of small nucleotides and nucleic acids is an efficient therapy for different types of cancers. The computational methods and tools deal with novel hypotheses of material discovery and design, improving properties and prediction of drug selection that could increase performance and reduce the time from drug discovery to marketing. This chapter presents a complete review of nanoinformatics databases, ontologies, machine learning, and its applications. There is a prospect in the field of nanobiotechnology to compare and integrate information across diverse fields of study through bioinformatics. The study includes the use of informatics to advance the biological and clinical applications of basic research in nanoscience and nanotechnology.

11.1 INTRODUCTION

In 1979, during a conference at the California Institute of Technology, Feynman's talk on, "There is Plenty of Room at the Bottom," was considered to be the theoretical starting point for the development of nanotechnology, suggesting that individual atoms and molecules could be manipulated, allowing the controlled production of materials at the nanometer scale with promising technical, industrial, and biological applications [21]. It brings revolutionary changes, not only in the fields of physics, chemistry, materials science, engineering, environmental sensing, manufacturing, and quantum computing, but also with across-the-board applications in clinical research and biotechnology. Consequentially, nanotechnology got recognition as an interdisciplinary field. Very rapidly in the eon of the 20th century, the different branches of science and technology started to coalesce under an umbrella of nanoscience and nanotechnology with a prefix of *nano*. The prefix may be used to show the utilization of tools of nanotechnology, nanomaterials, or sometimes to analyze the smallest creation of nature. An enormous amount of experimental work is being undertaken globally, either in the development of new techniques or in the development of new materials [54].

Medicine and the drug delivery system can be more effective in the near future through the application of nanotechnology, and hold great potential for the welfare of the society at large. This has been proven right with success in recent laboratory events on the usage of nanoparticles to deliver vaccines, delivery of chemotherapy drugs directly to cancer cells, etc. which have shown positive results in clinical trials [8, 9, 17]. These researches have led to a large variety of challenges, as well as opened up many opportunities and thus have given birth to various fields of nanotechnology like nanobiotechnology, nanobioinformatics, bionanoginomics, etc. An enormous amount of data is being generated in respect of nanomaterials on its synthesis, the study of its physical-chemical properties, in-vitro, and in-vivo testing, the study of toxicology, the effect of dosage, sequencing of genes, etc. and are in scattered forms of various research publications. The specific features determined by the interdisciplinary character and rapid evolution of this knowledge area are summarized for the data on the properties of nanosized objects. Thus, imposing challenges dealing with informatics like integrating and managing diverse information, classifying the types of nanomaterials, defining nomenclatures, taxonomies, the study of various simulation techniques and modeling techniques for nanoparticles (NPs), etc. This led to the evolution of new discipline to cope up and deal with the critical issue of management of innumerable large information and is called *nanoinformatics*. Nanoinformatics is a newly emerging field that involves a combination of methods and tools for dissemination of information on nanomaterials and the instruments and technologies involved in it.

The term "nanoinformatics" was coined in the year 2010, when a working group of experts Nanoinformatics 2010 was convened to formulate its objectives and directions of activity. The term has been generated by the needs of nano and information technologies, and the group was required to take up the interdisciplinary character (Figure 11.1) of this knowledge area and the permanent extension of its definition that reflect the appearance of new materials, devices, and applications into account [3]. The most comprehensive and detailed plan has been formulated under "Nanoinformatics Road Map 2020," which was adopted in the 2015 Nanoinformatics Seminar [24].

Many strategies have been applied to advance nanotechnology, based on the organization, interpretation, and prediction of the structure and physical-chemical properties of NPs and nanomaterials, which is gaining impetus. However, the applications of computer technologies, information science, and molecular simulations are enormously used to formulate methodologies research in the area of nanobiotechnology and nanoinformatics. These methodologies are suitable to produce qualitative concepts, insights, and

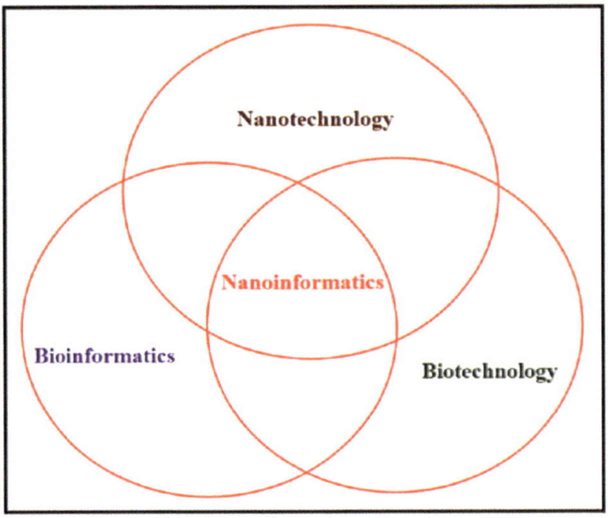

FIGURE 11.1 Nanoinformatics is a newly emerging informatics area, working at the intersection between bioinformatics, nanotechnology, and biotechnology.

design suggestions. While bioinformatics is frequently related to the use of computational tools to analyze DNA and protein sequence data, nanoinformatics is applied in the context of characterizing particles and materials with application in nano- and biotechnology, by modeling and simulating them, in many instances at the atomic level, using computational chemistry strategies [29]. Therefore, nanoinformatics cannot merely be considered as an application of informatics to nanotechnologies, but is involuntarily a mediator between two areas, is required to obtain deep insight towards the scientific essence of problems to look beyond external attributes such as database and ontologies [2]. Thus, it is more suitable to describe it as *"the science and practice of determining which information is relevant to the nanoscale science and engineering community, and then developing and implementing effective mechanisms for collecting, validating, storing, sharing, analyzing, modeling, and applying that information"* [18].

The newly emerging field nanoinformatics encompasses information of all nano-engineered materials. Nanoinformatics (Nanoinfo) coalesces the methods and tools for the ratification of data on nanomaterials as well as the instruments and technologies based on them (Figure 11.2).

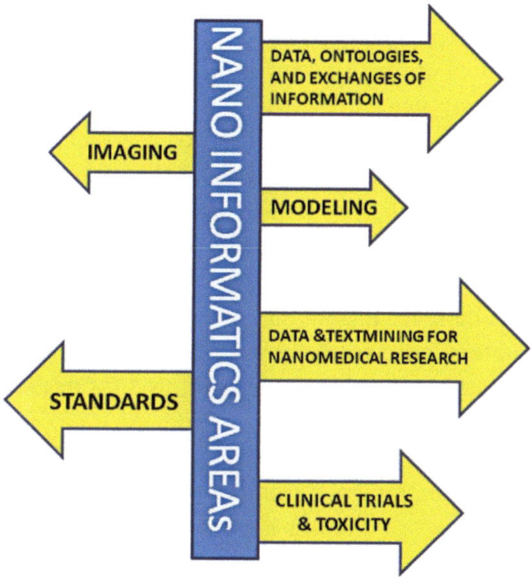

FIGURE 11.2 Different areas of nanoinformatics.

Pharmaceutical industries have been significantly active in the development of a unique drug molecule for various types of cancers in past decades. Target identification and validation of a drug are the prime and complex procedures in the design of a drug. Ligand molecules are designed to interact with the protein structure, and a few genes are also involved in processing the drug *in-vivo*. Still, there is a constant failure perceived in this field due to the complex mechanism of drug response and numerous factors that are involved in the key outcome of a drug [13]. Though the breakdown of a drug in the treatment of cancer comprises of various aspects, a large provenance towards the failure maybe because of the abnormalities at the chromosomal level [36]. To overcome this problem, nanoinformatics analysis plays a crucial role towards the discovery of effective nano-drugs to treat cancer.

Nanoinformatics plays a significant role in exploiting genomic, transcriptomic, and proteomic data to improve insights into the molecular mechanisms that underlie disease and search for drug targets that have novel mechanisms of action to accomplish unmet medical needs. The approach of studying the genetical basis of the disease is undergoing a revolution [35]. Of late, scientists have been exploring the substantial components of the expressed genome rather than focusing on individual genes. The wealth of molecular information being generated from the laboratories is continuously growing, and the same is the case with the volume of data pertaining to patient records. The crucial task is to find the ultimate ways to integrate these data to develop a novel insight into the genetics of cancers. The process of inventing any novel drug to treat any cancerous condition necessitates the focus of genetic variations and understanding of data analysis. In the case of the fusion gene, a modification at the nucleotide level has a significant influence on the protein [16]. To solve this problem, enhanced knowledge regarding the genes and their regulatory regions might provide a detailed basis for understanding cancer at the molecular level.

By discussing some of the aspects of the nanoinformatics, i.e., in silico methods, relevant databases, and ontologies, and by inculcating the author's original work, the chapter is penned to describe the recent trend of nanoinformatics for cancer therapeutic.

11.2 ONLINE RESOURCES (NANOINFORMATICS DATABASES AND ONTOLOGIES)

Researchers try to extract, confirm, or discover new knowledge from large databases by using data mining approaches. It was earlier used in large medical and biological databases to generate or confirm the research proposition. Their application has been fundamental for extracting knowledge patterns; for example, decision trees or association rules—either for descriptive or predictive tasks. Most of these tasks are based on inductive approaches. In order to predict the potential effects of the engineered nanomaterials (ENMs) based on their chemical properties, their mechanisms of action and the biological pathways, various data mining techniques with respect to nanobiology [46]. Biostatisticians and epidemiologists can monitor the impact of nanomedical applications by carrying out surveillance studies of nano-enabled drugs to monitor their potential toxicity as well as effectiveness in clinical practice, from the point of view of public health [15, 47]. In the era of personalized medicine—usually referred to as "genomic medicine," even though nanomedicine is already transforming this characterization—professionals and researchers have to study and assess the use of nanomedical treatments for specific patients based on individual features.

Development of in silico methods relies on good-quality experimental data on engineered nanomaterial toxicity as the set of parameters that determine the toxic potential of each type of ENMs in specific test species/taxa is largely unknown [19]. The need for data sharing and the complexities involved in the development of data standards for the application of nanotechnology in biomedicine is exemplified in several scientific scenarios. The ubiquitous databases and ontologies on the properties of nanomaterials are presented here.

11.2.1 DATABASES

11.2.1.1 CANANOLAB

The National Cancer Institute (NCI) Alliance for Nanotechnology in Cancer (Alliance), initiated in 2005, is a biomedical nanotechnology program supported by the NCI Office of Cancer Nanotechnology Research (OCNR). caNanoLab provides access to experimental and literature curated data from the NCI Nanotechnology Characterization Laboratory, the Alliance, and the greater cancer nanotechnology community. caNanoLab is a data-sharing portal designed to facilitate information sharing across the international biomedical nanotechnology research community to accelerate and corroborate the use of nanotechnology in biomedicine. caNanoLab provides support for the annotation of nanomaterials with characterizations resulting from physicochemical, *in vitro*, and *in vivo* assays and the sharing of these characterizations and associated nanotechnology protocols in a secure fashion [26].

The caNanoLab collaborative environment promotes data sharing and analysis across the cancer nanotechnology community to expedite and validate its use in biomedicine. caNanoLab is a web-based portal and data repository that allows researchers to submit and retrieve information on NPs including their composition, function (e.g., therapeutic, targeting, diagnostic imaging), physical (e.g., size, molecular weight) and *in vitro* (e.g., cytotoxicity, immunotoxicity) experimental characterizations. A given NPs entry also includes information on the protocols used for these characterizations and any related publications. Web-based forms are available to facilitate data submission, and submitters can customize the visibility of their data to range from private, to share with specified collaboration groups, to fully public.

11.2.1.2 ISA-TAB-NANO

The development of recent databases such as the Investigation Study Assay (ISA) tab-delimited (TAB) format (ISA-TAB-Nano) [49, 57] was established for the open data sharing and nano-data discovery and extraction. The ISA-TAB-Nano file arrangement proposes a common framework to record and incorporate nanomaterials descriptions and uses the ISA-TAB open-source framework file format. ISA-TAB-Nano describes four file formats for the sharing of the data:

i. The investigation file contains reference data about each investigation, study, assay, and protocol.
ii. The dtudy file provides the names and attributes of procedures applied for preparing samples for examination.
iii. The assay file is created for all assays performed, which define the values of measured endpoint variables and reference files for each examined material. For instance, physicochemical characterization of material (size, shape, etc.) and toxicological data (MTT, XTT, CFU assays and IC50, LD50 values) [28] and biomarkers increase value relative to control.
iv. The material file represents the nanomaterial and its structural and chemical components. For instance, different types of surface-coated metal oxides, dendrimers, carbon nanotubes (CNTs) that may contain diverse components like core and shell.

11.2.1.3 NANOE-TOX

The NanoE-TOX [32] is a database of the ecotoxicity of nanomaterials. The constant production and use of nanomaterials are hazardous to ecology. The in-vivo toxicity assays are ethically problematic and generate adverse environmental effects. This database generated by keyword search in the Thomson Reuters WoS database and retrieved existing ecotoxicological information of nanomaterials. The nanoecotoxicological data of commercially relevant eight NPs and its composites (CNTs, fullerenes, silver (Ag), titanium dioxide (TiO_2), zinc oxide (ZnO), cerium dioxide (CeO_2), copper oxide (CuO), and iron oxide (FeO_x, Fe_2O_3, Fe_3O_4)) are recorded in these databases for seven organism groups representing diverse trophic levels (bacteria, algae, crustaceans, ciliates, fish, yeasts, and nematodes) [33]. The database contains the comparative toxicity analysis of nanomaterials across the given species. NanoE-Tox database comprises significant data for nanomaterials' ecological hazard estimation and construction of models for predicting the toxic potential of nanomaterials.

11.2.2 ONTOLOGIES

11.2.2.1 ENANOMAPPER

eNanoMapper is the computational arrangement for toxicological data management of ENMs set by open models, ontologies, and an interoperable site to facilitate an effective, integrated method to research in nanotechnology. This program encourages the assessment of nanomaterials by building a modular, extensible infrastructure for open data sharing, data analysis, and the creation of computational toxicology models for nanomaterials. The eNanoMapper develops tools and models for a scientifically effectual evaluation of nanomaterials that support the idea of modern competent and environment-friendly nanomaterials and evaluation of existing materials. The nano-lazar application is developed under eNanoMapper project. The lazar (Lazy Structure-Activity Relationships) takes a chemical structure as a query and predicts the variety of toxic properties. The lazar uses an automated and reproducible read across the procedure to calculate predictions. Foundations for predictions, applicability domain estimations, and validation outcomes are presented in a clear graphical interface for the critical assessment by toxicological experts. Lazar is built on top of the OpenTox framework. It creates a community framework to collaborations, supports safe by designs ENMs, accelerates knowledge exchange through ontologies. The eNanoMapper ontology covers the full scope of terminology needed to support research into nanomaterial safety. It builds on multiple pre-existing external ontologies, such as the nanoparticle ontology (NPO) [56].

11.2.2.2 NANOPARTICLE ONTOLOGY

The NPO [57] was established to characterize nanomaterials involved in cancer research and maintained by the National Centre for Biomedical Ontology. The NPO follows the framework of the basic formal ontology (BFO) and achieved in the ontology web language (OWL), working on well-defined ontology design principles. This ontology represents information concerning preparation, chemical composition, and characterization of nanomaterials reported for different cancers. The NPO provides a common terminology and integration of data with the logical structure of fetching the data from tools and software related to cancer nanomedicine.

11.3 AMALGAMATION OF BIOINFORMATICS/ NANOINFORMATICS METHODS AND DRUG DISCOVERY

Developments in nanomedicine in the upcoming years will have diverse usage [42]. Nanotechnology has already taken a paradigm shift from being used as passive structures to active molecules as nanocarrier based targeted drug therapies or "smart drugs." These new drug therapies have shown to cause lesser side effects and more efficacy than traditional therapies [50]. In therapeutic medicine, nanotechnology is at the crossroad of both targeted drug delivery and core therapies. For instance, NPs can be injected into the tumor and then activated to generate heat and destroy cancer cells locally either by magnetic fields [44], x-rays, or light. Meanwhile, the encapsulation of current chemotherapy drugs or genes facilitates much more localized delivery, both reducing significantly the quantity of drugs absorbed by the patient for equal impact and the side effects on healthy tissues in the body [34]. However, a careful validation of its entrapment within target and efflux out of target is required prior to any direct therapeutic use. Therefore, nanoinformatics plays an extremely crucial and central role in order to figure out the exact mechanism underlying all the factors mentioned above and induces the cost-effective process of bringing the novel drug in the market. To understand the value of nanodelivery devices in modern medicine, we should consider that an effective drug must satisfy many requirements, including being of low toxicity, being efficacious, specific, and soluble, before it can be considered a viable candidate for clinical trials. In this section, we present a review of specificity, applicability, and integration of methodologies, software, computational tools of nanobioinformatics.

11.3.1 MOLECULAR MODELING AND SIMULATION

Computation tools have given some hopes for studying the protein-NPs complexes, which are very difficult to study experimentally. The study of protein-NPs can be studied computationally due to the availability of structural models of carbon nanomaterials such as CNTs. Fullerenols, which are derivatives of fullerenes, are currently in trial for diagnostic and therapeutic uses, although insufficient information is available about the structural interactions and toxicity of fullerenols in biosystems [51, 62] computationally studied the interactions in the fullerenol-lysozyme complex and compared the computational results with experimental results.

The investigation of nucleic acids and many protein structures has been done by crystallography, nuclear magnetic resonance, electron microscopy, and many other techniques [52]. Structural biology provides information on the static structures of biomolecules. However, in reality, biomolecules are highly dynamic, and their motion is important to their function. Various experimental techniques are available to help study the dynamics of biomolecules [41]. Computational power continues to increase, and the development of new theoretical methods offers hope of solving scientific problems at the molecular level. All the theoretical methods and computational techniques that are used to model the behavior of molecules are defined as molecular modeling.

11.3.2 PREDICTION OF ONCOGENIC PROTEIN STRUCTURE

The chromosomal rearrangements coding chimeric fusion proteins establish one of the most common mechanisms underlying malignant transformation in humans. These chimeric fusions are known as oncogenic fusion proteins (OFPs). The OFPs are proteins formed through the breakage and re-joining [38] of two or more genes that formerly coded for distinct proteins. The translation of these fusion gene consequences in a single or numerous polypeptides with functional properties derived from each of the unique proteins. Some of the recombinant fusion proteins formed artificially by recombinant DNA technology for use in research and therapeutics. The OFPs frequently entitle hybrid proteins made of polypeptides having diverse functions. The OFPs arise indeed of chromosomal rearrangements like chromosomal translocation, duplication, or retro-transposition that formulates a novel coding sequence containing parts of the coding sequences from two different genes. For instance, the occurrence of genomic rearrangements in mixed-lineage leukemia (MLL) or infant's leukemia is associated with poor prognosis, so the novel therapeutic strategies required exploring for this category of malignancy. To resolve these problems, the development of the three-dimensional molecular structure of MLL fusion protein required with superior importance to discover novel drug-like compounds that precisely target fusion protein. The author claims that the MLL fusion protein is modeled for the first time.

The genes originally code for separate proteins. The fusion proteins or chimeric proteins formulate through the joining of two or more genes that generate a fusion gene or oncogene. Translation of this fusion gene results in a single polypeptide with functional properties derived from each of the original proteins. Chimeric proteins occur naturally when a large-scale mutation or chromosomal translocation produces a novel coding sequence containing components of the coding sequences from two different genes. Naturally, occurring fusion proteins are important in cancer, where they may function as oncoproteins. The fusion gene can be transcribed, spliced, and translated to produce a functional fusion protein. Many important cancer-promoting oncogenes are fusion genes produced in this way. For instance, the AF9 gene from chromosome 9 joints with the MLL gene on chromosome 11, to form MLL-AF9 fusion gene (Figure 11.3). Amongst all MLL translocations, around 50% of infant AML cases comprises of t (9, 11) (p22, q23) rearrangement. AF9 gene, also known as LTG9 or MLLT3, is located at short arm p22 of chromosome 9 [30, 43, 61]. The accumulating evidence suggests that leukemogenesis caused by the formation of the MLL-AF9 fusion protein through the mechanism of these partner genes is unsigned. In contrast, few *in-vitro* and *in-vivo* examination revealed that MLL-AF9 alters myeloid progenitor cells and suppresses specific HOX genes. The mice with knock-in MLL-AF9 fusion genes demonstrated the abnormal proliferation of hematopoietic cells and developed AML identical to the patient with t (9; 11) translocation [12, 20, 31]. Additional to this, MLL, and AF9 wild protein participates indispensably during the hematopoiesis/embryogenesis process and are elements of protein complexes resulting in target gene transcriptional initiation (MLL) and elongation (AF9).

Therefore, it was hypothesized that MLL-AF9 fusion combines these characteristics, resulting in increased activation of target genes, which may be interrupted hematopoietic cell differentiation and ultimately leads to leukemogenesis [11, 48, 53].

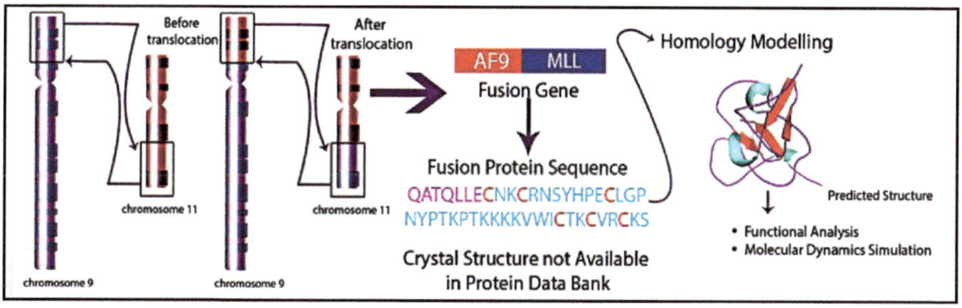

FIGURE 11.3 Reciprocal translocation between chromosome 9 and 11 leads to AF9-MLL fusion gene codes for Novel Fusion Protein. Structure prediction of fusion protein through homology modeling [16].

Source: Reprinted from Ref. [16]. https://innovareacademics.in/journals/index.php/ijpps/article/view/8721. https://creativecommons.org/licenses/by/4.0/

Native atomic bonds and interactions to drug binding sites mainly express the mechanisms and effectiveness of therapeutic action. The protein structure prediction is accurately comparable to experimental residues is challenging [55]. The critical assessment of protein structure prediction (CASP) community provides an opportunity to research groups through objectively test their structure prediction methods. It presents an independent evaluation of the cutting- edge protein structure modeling to the research community and software users. The variety of methods generates protein models. Though it is unclear what exactly done, the methods roughly classified the following groups: "Modeler," "Raptor," "Zhang" group, "Rosetta," "Lee" group, and others. The use of physics-based methods in protein structure refinement is highly encouraging. It is highly complementary to knowledge-based methods and allows the future repetitive refinement of protein structures to experimental accuracies. In this study, MD simulations and loop refinements further polished predicted models of fusion proteins [25] in CASP11 recommend that the refinement performs significant dynamics, where the averaged models have very poor MolProbity scores. Molecular dynamics

(MD) simulations used in combination with a better-quality selection and averaging protocol. The preliminary 3D model of AF9-MLL fusion protein acquired from homology modeling and further refined by MD simulation to improve the accuracy of the structure.

Homology models are erroneous as structure emerges by a course of amino acid insertions, substitution, and deletions [1, 6, 7]. Imprecision in model comprises of deformation in secondary structure elements, side-chain packaging error, and inadequately delineated loop conformations, which necessitates that all predicted structures are mandatory for further refinement. The Ramachandran plot analysis performed using PROCHECK, as demonstrated in Figure 11.4a. The Ramachandran plot of predicted and refined models indicated φ and ψ angle in the most favored regions, allowed regions and outliner regions (Table 11.1). Consistency of the generated model further computed by ERRAT, verify 3D, and PROVE. This is suggestive of the fact that the model so constructed is of the superior kind (Figure 11.4b, c, and d).

TABLE 11.1 Comparative Values of Procheck, Errat, Verify_3D, Prove in Different Stages of Refinement Used in I-TASSER Software [16]

Validation		Predicted Model	Model Energy Minimized	Model Refined
PROCHECK	Regions of Ramachandran Plot			
	Favored	46.9%	67.2%	86.5%
	Additionally allowed	43.8%	27.1%	10.4%
	Generously allowed	6.2%	2.1%	2.1%
	Disallowed	3.1%	3.1%	1.0%
ERRAT		86.86	95.91	77.27
VERIFY_3D		82.24	90.65	72.90
PROVE Z score		Error	0.76	0.54

Source: Reprinted from Ref. [16]. https://innovareacademics.in/journals/index.php/ijpps/article/view/8721. https://creativecommons.org/licenses/by/4.0/

11.3.3 DOCKING AND SIMULATION STUDIES AND SYNTHETIC DNA AS NANO DRUG

Developments in nanotechnology are opening up the prospects for nanomedicine and regenerative medicine where informatics and DNA

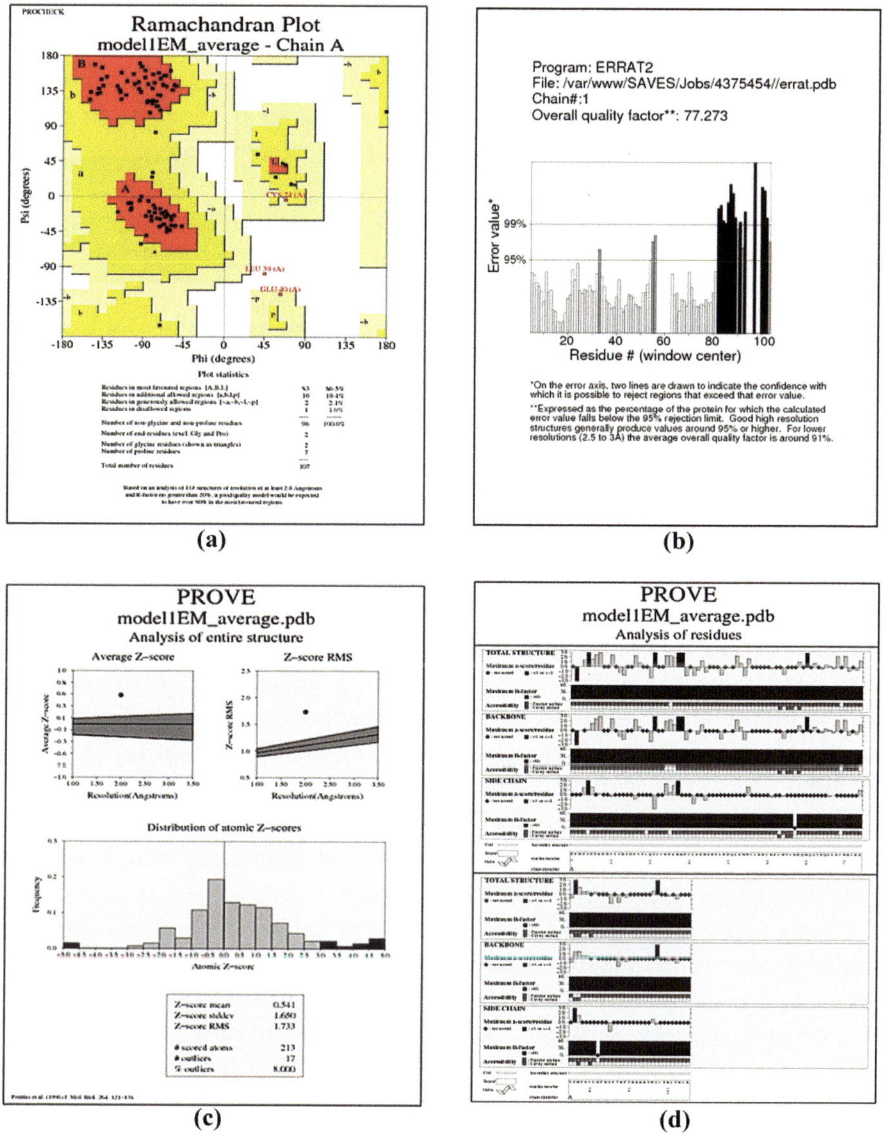

FIGURE 11.4 Structural quality assessment of modeled AF9-MLL fusion protein. (a) Ramachandran plot values showing a number of residues in favored, allowed, and outlier regions. (b) Errat plot where black bars show the misfolded region, gray bars demonstrate the error region between 95% and 99%, and white bars indicate the region having less error rate for protein folding. (c and d) PROVE shows Z-score with respect to a residue with insignificant error standards of individual amino acid residues in modeled fusion protein. [16].

computing can become the catalysts enabling health care applications at sub-molecular or atomic scales. Nanoinformatics can accelerate the introduction of nano-related research and applications into clinical practice, leading to an area that could be called "translational nanoinformatics." At the same time, DNA, and RNA computing presents an entirely novel paradigm for computation [58].

In this section, the author discusses their work on synthetic ssDNA as a drug for OFP. The functional analysis revealed that AF9-MLL OFP possesses the CXXC domain: CpG DNA binding domain. The protein-DNA interactions are the foundation of gene expression and DNA modification. These interactions are the prerequisite for the differentiating structural models. As only a few numbers of structures for protein-DNA complexes determined by experimental methods, computation methods postulate a promising way to fulfill the need. The vertebrate DNA is able to chemically alter by methylation of the five locations of the cytosine base in the context of CpG (5'-Cytosine-phosphate-Guanine-3') dinucleotides. This alteration generates a binding site for the methyl-CpG-binding domain (MBD) proteins that direct the chromatin-modifying actions that considered contributing to transcriptional repression and control heterochromatic regions of the genome. In contrast with DNA methylation that found broadly across vertebrate genomes, non-methylated DNA is concentrated in regions known as CGIs (CpG islands). Recently, families of proteins which encode a ZF-CXXC (zinc finger-CXXC) domain have been determined to specifically distinguish non-methylated DNA and recruit chromatin-modifying activities to CGI elements [37].

The functional analysis and active site predictions of our study disclose that all fusion protein contains CXXC domain. The CXXC protein domain present in normal MLL also retains in MLL [16] OFPs, which bind to nonmethylated CpG DNA [4, 5, 45], and this interaction is essential for the recruitment of MLL to Hoxa9 and leukemogenesis.

To our best of knowledge, the study is the first in-silico report on the significant association of synthetic CpG DNA as the newer alternative drug like compound for MLL fusions to target CXXC domain [45]. In addition, we tried to define the significance of CXXC domain in AF9-MLL fusion protein through in-silico prediction and MD simulation methods. Molecular docking and MD simulations performed to study the potential of bonding, optimum-binding scores, nature of interactions, and thermodynamic stability.

The ZF-CXXC domain recognizes only non-methylated CpG motif. The published literature insinuates that the domain does not distinguish DNA bases beyond the CpG dinucleotides [23]. For docking studies, the active sites of CXXC domain (PDB id: 2jyi) affirmed by the literature survey, the common active site amino acids are cys1170, thr1171, asn1172, cys1173, leu1174, asp1175, lys1176, lys1185, lys1186, gln1187, cys1188, cys1189, and lys1190. The protein structure subjected to simulation. After refinement, the post docking reverted to check binding efficacy. Table 11.2 shows cys1188 have the lowest score amongst another residue. In certain cases, protein movements are limited, and the ligand fits into a static binding pocket as a key fits into a lock. Computational simulation of protein affects the overall stability and alterations in ligand binding properties. After the simulation, a single protein conformation expresses about protein dynamics. The post-docking results indicate the change in binding energy (Table 11.2).

The docking performed to determine the binding efficacy of CpG DNA and CXXC domain. Since the interacting residues and H-bonds involved in the DNA-protein interaction have a significant role in governing the binding affinity, it is important to analyze the difference in the interaction patterns of CpG DNA with both wild type and mutant CXXC domain. The docking result and ligand interaction plot (Figure 11.5a) illustrate that CpG DNA tends to interact with six connecting amino acids. The hydrogen bonds are cys1188 (3.11 Å, 3.15 Å, 2.76 Å), arg1154 (2.49 Å, 2.79 Å, 2.84 Å, 3.87 Å), arg1150 (2.36 Å, 2.65 Å), lys1176 (2.84 Å), cys1189 (3.25 Å), asp1166 (2.49 Å). The ligand interaction plot (Figure 11.5) reveals the presence of 12 hydrogen bonds and binding free energy score –21.20 kcal/mol (Table 11.2) with residue cys1188 before simulation.

However, recent studies suggest that point mutation does not favor protein-DNA binding, thus, influences the binding energy of CpG DNA with CXXC domain. The *in-silico* point mutation of cysteine to aspartic acid (C1188D) computed in MVD software.

Table 11.2 demonstrates the higher docking score of 2.26 kcal/mol of mutant before simulation with only three H-bonds. In contrast, after simulation, the docking score displays lower energy (–4.0 kcal/mol) for the mutant. The ligand interaction plot (Figure 11.6) for wild-type and mutant after simulation shows the intermolecular H-bonding between the protein and CpG DNA that plays a vital role in stabilizing the protein-DNA complexes. The post-simulation interacting residue with H-bond distance for wild type are asn1151 (2.84 Å, 2.83 Å), lys1176 (2.83 Å), asn1183 (3.06 Å), gln1187 (3.06 Å), cys1188 (2.77 Å), cys1194 (3.07 Å, 2.47 Å) and with mutant are asn1183 (2.32 Å), gln (2.80 Å, 2.66 Å), arg1192 (2.57 Å, 2.77 Å), and cys1194 (2.66 Å).

TABLE 11.2 Molecular Docking Studies Performed for CXXC Domain Mutant Type Against CpG DNA

	Binding Energy (kcal/mol)	
Residue	**Before simulation**	**After simulation**
Cys1188 (wild type-CXXC domain)	–21.2	–12.2
C1188D (mutant)	2.2	–4.0

The results suggest that there is a net shrink in H-bond in wild type protein (see Figures 11.5a and 11.6a) and rise in H-bond in the mutant (see Figures 11.5b and 11.6b) after simulation. For the reason that after MD simulation, the structure is proximal to the in-vitro environment presenting an altered docking score for amino acid residues. This computational investigation strongly correlates with the in-vitro consequence [10].

Modification in amino acid residue feasibly perturbs hydrogen bonding and salt bridge unsettling structural stability and binding. Upon mutation of cysteine to aspartic acid experienced a large conformational change relative to its positioning when protein bound to CpG DNA. Amongst all DNA binding domains in vertebrates, only the CXXC domain discriminates non-methylated CpG DNA that leads to the uncontrolled proliferation of cells in MLL. The mutation leads to alter the stability of the protein. The physicochemical characteristics of mutated amino acid differ from wild type, causing transformation.

Our examination on the contact of mutation revealed that the change in size, charge, and hydrophobicity might lead to loss of interaction with other molecules or residues, which in turn could result in adverse effects. Taking into consideration of the parameters such as RMSD and RMSF, MD simulation was performed for 10 ns with a view to observe the change (if any) in conformation behavior of the mutant protein C1188D, as compared to that of the native (wild type CXXC domain). The MD simulation on AF9-MLL fusion protein for 10 ns measure up to examine the dynamic performance of CXXC domain retains in this fusion protein.

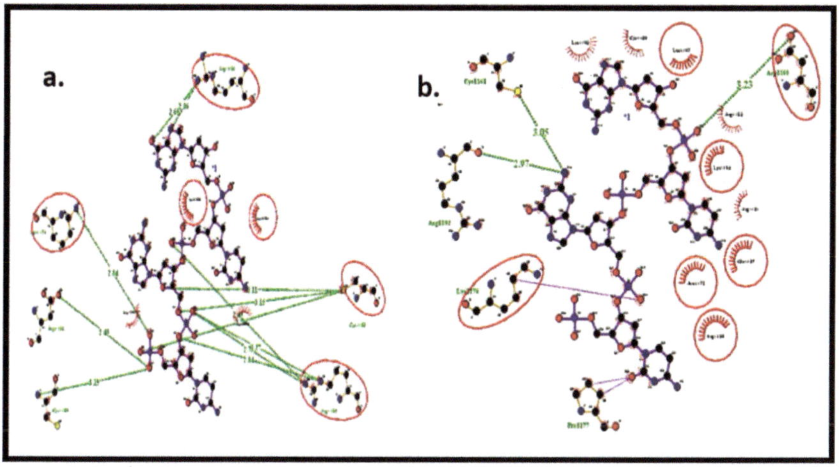

FIGURE 11.5 Protein DNA interaction map before MD simulation: (a) Interaction of CpG DNA with wild type CXXC domain, and (b) Interaction of CpG DNA with point mutation cysteine 1188 to aspartic acid 1188.

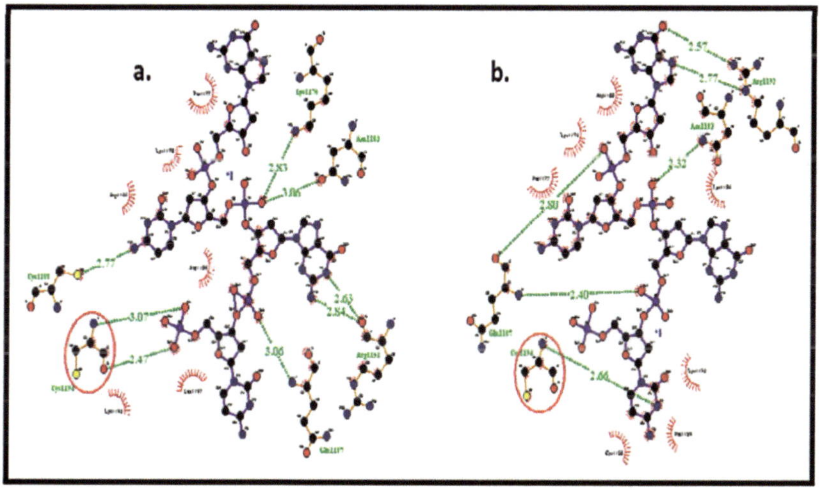

FIGURE 11.6 Protein DNA interaction map after MD simulation (a) Interaction of CpG DNA with wild type CXXC domain and (b) Interaction of CpG DNA with point mutation cysteine 1188 to aspartic acid 1188.

All protein structures experienced initial fluctuation due to the kinetic shock applied during the MD simulation. Interestingly, the native and the mutant proteins showed different types of deviation throughout the simulation. Overall, RMSD ranges of ~0.1–0.8 and ~0.2–0.45 nm perceived for the native and the fusion protein respectively (Figures 11.7 and 11.8). This magnitude of fluctuation, together with the difference between the average RMSD values after the relaxation period (1 ns), suggested that simulation produced stable trajectories, thus providing a suitable basis for further analyses (Figures 11.8).

The RMSF values of the Cα atom of native and mutant protein were calculated with the aim of determining whether mutation and fusion affected the dynamic behavior of each residue in the domain (Figures 11.7 and 11.8). The analysis unveiled the existence of a degree of flexibility in mutant structure as compared to that of the wild-type CXXC domain. The wild-type and mutant domain attained a high level of fluctuation in the residue positions 1165–1175 and 1182–1188. The residue 1188 manifests fluctuation up to 0.4 nm. The fusion protein residues 40, 62, and 85 determined maximum variation. Moreover, interestingly, the amino acid residues present in the fusion protein exhibited high fluctuation, indicating that the fusion has affected the protein conformation leading to an increased flexibility of residues in some regions.

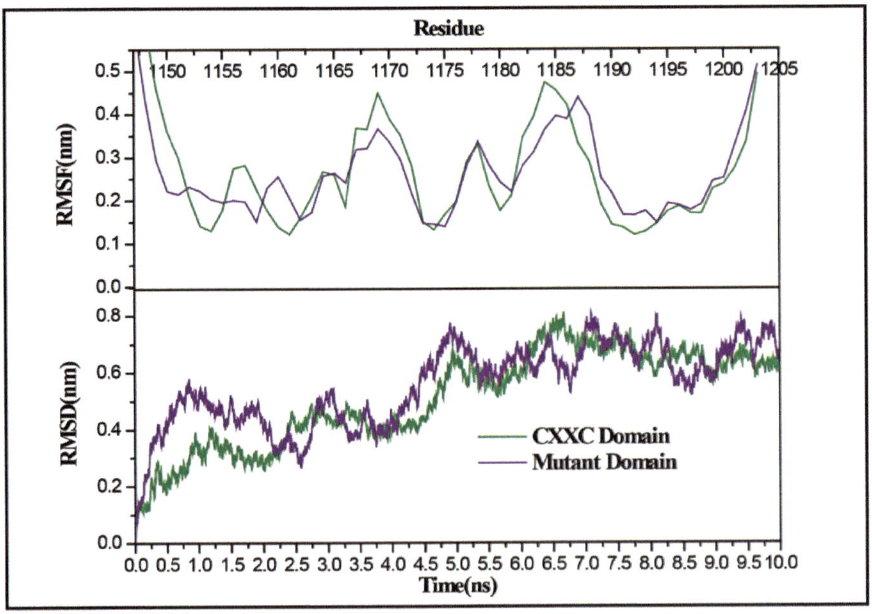

FIGURE 11.7 RMSD and RMSF trajectory of the CXXC domain wild type, mutant backbone during the 10 ns long MD simulations.

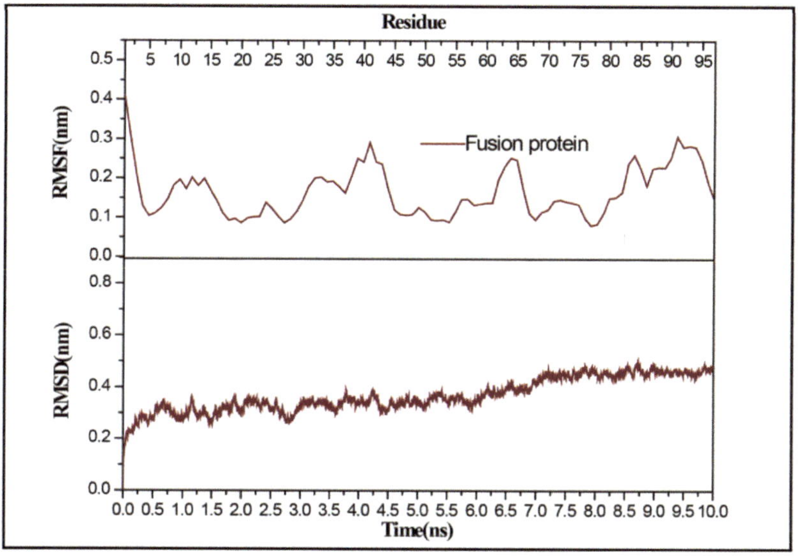

FIGURE 11.8 RMSD and RMSF of AF9-MLL fusion protein complexes during the MD simulations.

The radius of gyration (Rg) was applied to measure and understand the compactness of the protein complexes (Figure 11.9). The Rg value of ~1.3 nm and ~1.5 nm was perceived for the native and mutant consequently. The molecular compactness was observed for the mutant domain that presents different deviation patterns and convergence. However, the fusion protein shows good convergence with a higher Rg value ~1.6 nm.

During this study, computational microscope MD analysis covers the use of force fields for atoms present in a macromolecule. This precise prediction helps us understand its motion and interprets the small difference that occurred due to a variation. The mutation in CXXC domain alters the hydrogen bond network around the CpG sites. There is a small reduction in the net number of protein-DNA hydrogen bonds, including a loss of a crucial hydrogen bond between the cytosine bases. Point mutations in the CXXC domain that suppress myeloid transformation by MLL fusion protein also abolished recognition and binding of non-methylated CpG DNA sites in-vitro and transactivation in-vivo [10]. Therefore, the induced mutation is an alternative of directing CXXC domain in MLL associated malignancies.

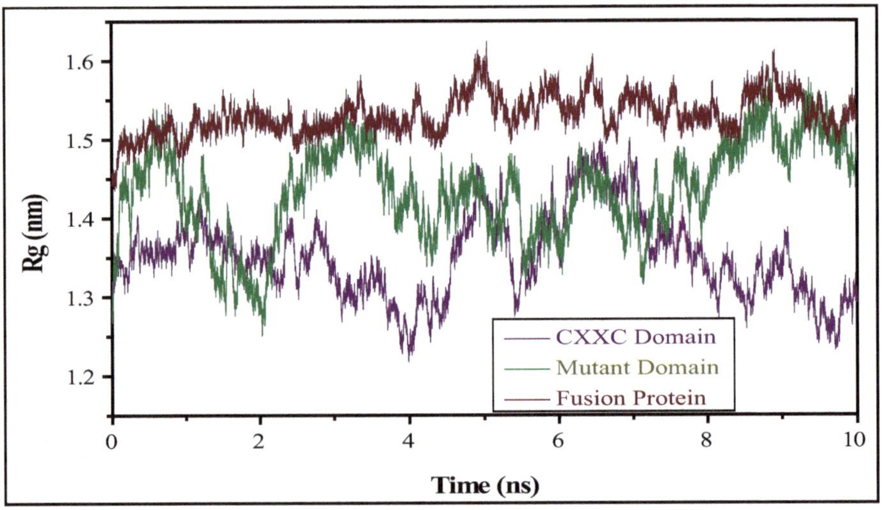

FIGURE 11.9 Radius of gyration (ROG) of CXXC domain.

In conclusion, the homology model was developed and validated for MLL fusions employing bioinformatics tools. These analyses validated that the simulated model is best, robust as well as reliable enough to be used for future study. Eventually, these molecular and structural studies result in the advancement of newer therapies.

In this study, virtual screening, molecular docking, and dynamics simulation applied to find the inhibitors of the oncogenic fusion protein, which may be potent in curing pediatric leukemia. The MD study also manifests the stability during the simulation process, which supports the virtual screening and docking results. However, further, in-vitro validations require establishing in-silico consequences.

11.4 CONCLUSIONS

The chapter describes about nanoinformatics playing a central role in the scientific inquiry, design, and technological development of NPs production. The links between the *in-silico, in-vitro,* and *in-vivo* examination can help in the diagnosis and therapy of cancer [60]. By building detailed models and *in-silico* computerized simulations, it is possible to design their structure and predict their properties. To understand the value of nano delivery devices in modern medicine, we should consider that an effective nano-drug must satisfy requirements of being low toxic, being efficacious, specific, soluble, and cost-effective before it can be considered a viable candidate for clinical trials. The computational data, strategies, and procedures must be designed in such a manner to integrate anti-microbial anti-cancer information for nano-applications. Linked to both simulation and clinical trial results, these nanoinformatics applications could help identify and compensate for individual reactions to these pathogens and support the needed clinical guidelines.

However, due to lack of information, there are many major research challenges which include the addition of new reference nanomaterials, characterization of improved nanomaterials, variation in the experiments, materials, and methodologies, error quantification and uncertainty in the methods and protocols used to produce the data, evaluation, and risk management.

ACKNOWLEDGMENT

Authors are thankful to Prof. S.P. Bhatnagar, Head, Department of Physics, and M.K. Bhavnagar University for continuous support and providing computational facility.

KEYWORDS

- **biomarkers**
- **bionanoginomics**
- **caNanoLab**
- **cancer therapeutics**
- **chimeric fusions**
- **chromosomal translocation**
- **computational chemistry**
- **database**
- **DNA**
- **ecotoxicity**
- **eNanoMapper**
- **genomic data**
- **genomic medicine**
- **in silico**
- **infant's leukemia**
- **in-vitro**
- **in-vivo**
- **ISA-TAB-nano**
- **I-TASSER software**

- **leukemogenesis**
- **methyl-CpG-binding domain (MBD)**
- **molecular dynamics (MD)**
- **molecular modeling**
- **MVD software**
- **nanobioinformatics**
- **nanobiotechnology**
- **NanoE-TOX**
- **nanoinformatics**
- **nano-lazar**
- **nanomedicine**
- **nanoscience**
- **nanotechnology**
- **oncogenic protein structure**
- **ontologies**
- **personalized medicine**
- **physcio-chemical**
- **polypeptides**
- **protein sequence**
- **proteomic data**
- **RNA**
- **simulation**
- **smart drugs**
- **transcriptomic data**

REFERENCES

1. Abagyan, R., Batalov, S., Cardozo, T., Totrov, M., Webber, J., & Zhou, Y., (1997). Homology modeling with internal coordinate mechanics: deformation zone mapping and improvements of models via conformational search. *Proteins, 29*(1), 29–37.
2. Agrawal, A., & Choudhary, A., (2016). Perspective: Materials informatics and big data: Realization of the fourth paradigm of science in materials science. *APL Materials, 4*, 053208.
3. Amherst, M., A., (2011). National nanomanufacturing network. *Nanomanufacturing*, 2020. Roadmap. http://eprints.internano.org/607/ (Accessed on 5 November 2019).
4. Armstrong, S. A., Jane, E. S., Lewis, B. S., Rob, P., Monique, L., Den Boer, M. D., Minden, S. E., Sallan, E. S. L., Todd, R. G., & Stanley, J. K., (2002). MLL translocations specify a distinct gene expression profile that distinguishes a unique leukemia. *Nature Genetics, 30*, 41–47.
5. Ayton, P. M., Everett, H. C., & Michael, L. C., (2004). Binding to nonmethylated CpG DNA is essential for target recognition, transactivation, and myeloid transformation by an MLL oncoprotein. *Molecular and Cellular Biology, 24*(23), 10470–10478.
6. Baker, D., & Andrej, S., (2001). Protein structure prediction and structural genomics. *Science, 294*, 5540, 93–96.
7. Bradley, P., Kira, M. S. M., & David, B., (2005). Toward high-resolution de novo structure prediction for small proteins. *Science, 309*(5742), 1868–1871.
8. Brannon-Peppas, L., & James, O. B., (2012). Nanoparticle and targeted systems for cancer therapy. *Advanced Drug Delivery Reviews, 64*, 206–212.

9. Brigger, I., Catherine, D., & Patrick, C., (2002). Nanoparticles in cancer therapy and diagnosis. *Advanced Drug Delivery Reviews, 13, 54*(5), 631–651.
10. Cierpicki, T., et al., (2010). Nature structural & molecular biology. *Structure of the MLL CXXC Domain-DNA Complex and its Functional Role in MLL-AF9 Leukemia, 17*(1), 62–68.
11. Collins, E. C., et al., (2002). Molecular and cellular biology. *Mouse Af9 is a Controller of Embryo Patterning, Like Mll, Whose Human Homologue Fuses with Af9 After Chromosomal Translocation in Leukemia, 22*(20), 7313–7324.
12. Corral, J. et al., (1996). Cell. *An Mll–AF9 Fusion Gene Made by Homologous Recombination Causes Acute Leukemia in Chimeric Mice: A Method to Create Fusion Oncogenes, 85*(6), 853–861.
13. Couch, R. D., & Bryan, T. M., (2012). Personalized medicine: Changing the paradigm of drug development. *Molecular Profiling: Methods and Protocols*, 367–378.
14. Daga, A., et al., (2016). Interdisciplinary sciences: Computational life sciences. *Significant Role of Segmental Duplications and SIDD Sites in Chromosomal Translocations of Hematological Malignancies: A Multi-Parametric Bioinformatic Analysis*, 1–9.
15. Daum, N., Tscheka, C., Neumeyer, A., & Schneider, M., (2012). Novel approaches for drug delivery systems in nanomedicine: Effects of particle design and shape. *Wiley Interdiscip. Rev. Nanomed. Nanobiotechnol., 4*(1), 52–65.
16. Dave, M., Aditi, D., & Rakesh, R., (2015). Structural and functional analysis of AF9-MLL oncogenic fusion protein using homology modeling and simulation based approach. *Int. J. Pharm. Pharm. Sci., 7*(12), 155–161.
17. Davis, M., E., & Dong, M., S., (2008). Nanoparticle therapeutics: An emerging treatment modality for cancer. *Nature Reviews Drug Discovery,* 7(9), 771.
18. Dieb, T., M., Yoshioka, M., Hara, S., & Newton, M. C., (2015). *Beilstein, J. Nanotechnol., 6,* 1872–1882. doi: 10.3762/bjnano.6.190.
19. Djurišić, A. B., Leung, Y. H., Ng, A. M., Xu, X. Y., Lee, P. K., Degger, N., & Wu, R. S., (2015). Small. *Toxicity of Metal Oxide Nanoparticles: Mechanisms, Characterization, and Avoiding Experimental Artefacts, 11*(1), 26–44.
20. Dobson, C. L., Warren, A. J., Pannell, R., Forster, A., Lavenir, I., Corral, J., et al., (2003). Oncogene. *The Mll-AF9 Gene Fusion in Mice Controls Myeloproliferation and Specifies Acute Myeloid AML Without Further Processing, 2,* 8448–8459.
21. Drexler, K. E., (1992). Nanotechnology: The past and the future. *Science, 255,* 268–269.
22. *eNanoMapper: Computational Infrastructure for Toxicology Data Management of Engineered Nanomaterials.* https://www.ebi.ac.uk/ols/ontologies/enm (Accessed on 5 November 2019).
23. Erfurth, F. E., et al., (2008). Proceedings of the national academy of sciences. *MLL Protects CpG Clusters from Methylation Within the Hoxa9 Gene, Maintaining Transcript Expression, 105*(21), 7517–7522.
24. Erkimbaev, A. O., Zitserman, V., Yu, K. G. A., & Trakhtengerts, M. S., (2016). Scientific and technical information processing. © *Allerton Press, Inc., 43*(4), 199–216.

25. Feig, M., & Vahid, M., (2016). Protein structure refinement via molecular-dynamics simulations: What works and what does not? *Proteins: Structure, Function, and Bioinformatics, 84*(S1), 282–292.
26. Gaheen, S., et al., (2013). caNanoLab: Data sharing to expedite the use of nanotechnology in biomedicine. *Computational Science and Discovery, 6*(1), 014010.
27. Gaheen, S., et al., (2013). caNanoLab: Data sharing to expedite the use of nanotechnology in biomedicine. *Computational Science and Discovery, 6*(1), 014010. https://cananolab.nci.nih.gov/caNanoLab/ (Accessed on 5 November 2019).
28. Golbamaki, N., Rasulev, B., Cassano, A., Marchese, R. R. L., Benfenati, E., Leszczynski, J., & Cronin, M. T., (2015). *Nanoscale, 7*(6), 2154–2198.
29. González, F., Tomás, P., Sergio, G., Daniela, A., G., Claudia, S., Alejandro, Y., Leonardo, S. S. V., Felipe, L., Hegaly, M., & Raúl, E. C., (2011). *Biol. Res., 44*, 43–51.
30. Iida, S., et al., (1993). Oncogene. *MLLT3 Gene on 9p22 Involved in t (9, 11) Leukemia Encodes a Serine/Proline Rich Protein Homologous to MLLT1 on 19,* p. 8, 11, 13, 3085–3092.
31. Joh, T., et al., (1999). Oncogene. *Establishment of an Inducible Expression System of Chimeric MLL-LTG9 Protein and Inhibition of Hox a7, Hox b7 and Hox c9 Expression by MLL-LTG9 in 32Dcl3 Cells, 18*(4), 1125–1130.
32. Juganson, K., et al., (2015). NanoE-Tox: New and in-depth database concerning ecotoxicity of nanomaterials. *Beilstein Journal of Nanotechnology, 6*, 1788.
33. Kahru, A. D., (2010). From ecotoxicology to nanoecotoxicology. *HC Toxicology, 269*(2&3), 105–119.
34. Koo, Yong-Eun, L., et al., (2007). Applied optics. *Photonic Explorers Based on Multifunctional Nanoplatforms for Biosensing and Photodynamic Therapy, 46*(10), 1924–1930.
35. Lamb, J., et al., (2006). The connectivity map: Using gene-expression signatures to connect small molecules, genes, and disease. *Science, 313*(5795), 1929–1935.
36. Lewis, & Lionel, D., (2005). Personalized drug therapy, the genome, the chip and the physician. *British Journal of Clinical Pharmacology, 60*(1), 1–4.
37. Long, H. K., Neil, P. B., & Robert, J. K., (2013). ZF-CxxC Domain-Containing Proteins, CpG Islands and the Chromatin Connection, *Biochem. Soc. Trans. 41*(3), 727–740.
38. Dave, M., & Rawal, R., (2013). Insilco Analysis of Translocation Breakpoint Region of MLL (Mixed Lineage- Leukemia), *3rd International Conference on Proteomics and Bioinformatics. 6*(7), 88.
39. Maojo, V., et al., (2010). Nanoinformatics and DNA-Based Computing: Catalyzing Nanomedicine, *Pediatric Research, 67*(5), 481.
40. Maojo, V., et al., (2012). Nanoinformatics: A new area of research in nanomedicine. *International Journal of Nanomedicine, 7*, 3867.
41. Martin, K., & Andrew, M. J., (2002). Molecular Dynamics Simulations of Biomolecules, *Nature Structural Biology, 9*, 646–652.
42. Mehta, R. V., (2017). Modern application of bioequivalence and bioavailability. *Impact of Magnetic Nanomaterials on Biotechnology and Biomedicine, 2*, 2.
43. Nakamura, T., et al., (1993). Proceedings of the National Academy of Sciences. *Genes on Chromosomes 4, 9, and 19 Involved in 11q23 Abnormalities in Acute Leukemia Share Sequence Homology and/or Common Motifs, 90*(10), 4631–4635.

44. Nidhi, A., Bhupendra, C., Mehta, R. V., & Upadhyay, R. V., (2011). Biodegradable thermoresponsive olymeric magnetic nanoparticles: a new drug delivery platform for doxorubicin, *Journal of Nanoparticle Research, 13*(4), 1677–1688.
45. Pandya, M., et al., (2016). Targeting MLL-CXXC Domain with Synthetic CpG Dinucleotides: Docking and Molecular Dynamics Simulation Based Approach. *Advanced Materials Research,* 1141.
46. Panneerselvam, S., & Sangdun, C., (2014). Nanoinformatics: Emerging databases and available tools. *International Journal of Molecular Sciences*, *15*(5), 7158–7182.
47. Pautler, M., & Brenner, S., (2010). Nanomedicine: Promises and challenges for the future of public health. *Int. J. Nanomedicine*, *5*, 803–809.
48. Pina, C., et al., (2008). Cell stem cell. *MLLT3 Regulates Early Human Erythroid and Megakaryocytic Cell Fate*, *2*(3), 264–273.
49. Rallo, R., (2014). An ISA-TAB nano compliant data management system for nanosafety modeling. *NanoWG Meeting.*
50. Ruiz-Hernandez, E., Alejandro, B., & María, V. R., (2011). *Smart Drug Delivery Through DNA/Magnetic Nanoparticle Gates*, *5*(2), 1259–1266.
51. Sachkova, A. S., et al., (2017). Biochemistry and biophysics reports. *On Mechanism of Antioxidant Effect of Fullerenols*, *9*, 1–8.
52. Silke, B., (2011). Preparation of functional magnetic nanocomposites and hybrid materials: recent progress and future directions. *Nanoscale*, *3*, 877.
53. Slany, & Robert, K., (2009). Haematologica. *The Molecular Biology of Mixed Lineage Leukemia, 94*(7), 984–993.
54. Som, C., Bernd, N., Harald, F. K., & Peter, W., (2012). Toward the development of decision supporting tools that can be used for safe production and use of nanomaterials. *Accounts of Chemical Research, 46*(3), 863–872.
55. Soundararajan, V., & Murali, A., (2014). Scientific reports. *Global Connectivity of Hub Residues in Oncoprotein Structures Encodes Genetic Factors Dictating Personalized Drug Response to Targeted Cancer Therapy*, *4*, 7294.
56. Thomas, D. G., Gaheen, S., Harper, S. L., Fritts, M., Klaessig, F., Hahn-Dantona, E., Paik, D., Pan, S., Stafford, G. A., Freund, E. T., et al., (2013). ISA-TAB-nano: A specification for sharing nanomaterial research data in spreadsheet-based format. *BMC Biotechnol., 13*, 2.
57. Thomas, D. G., Rohit, V. P., & Nathan, A. B., (2011). Nanoparticle ontology for cancer nanotechnology research. *Journal of Biomedical Informatics, 44*(1), 59–74.
58. Victor, M., Fernando, M. S., Casimir, K., Alfonso, R. P., & Martin, F., (2010). Pediatric research. *Nanoinformatics and DNA-Based Computing: Catalyzing Nanomedicine*, *67*, 5.
59. Vladimir, M., Stefan, E., Kai, S., Brian, F., Michelle, L., & Peter, K., (2009). Occupational safety and health in nanotechnology and Organization for Economic Cooperation and Development. *Journal of Nanoparticle Research*, *11*(7), 1587–1591.
60. Wu, Z., Chen, K., Yildiz, I., et al., (2012). Nanoscale. *Development of Viral Nanoparticles for Efficient Intracellular Delivery*, *4*, 3567–3576.
61. Yamamoto, K., et al., (1994). Blood. *A Reverse Transcriptase-Polymerase Chain Reaction Detects Heterogeneous Chimeric mRNAs in Leukemia's with 11q23 Abnormalities*, *83*(10), 2912–2921.

62. Yang, S. T., Wang, H., Guo, L., Gao, Y., Liu, Y., & Cao, A., (2008). Nanotechnology. *Interaction of Fullerenol with Lysozyme Investigated by Experimental and Computational Approaches*, *19*, 395101.

CHAPTER 12

Thin Films: An Overview

PANKAJ KUMAR MISHRA

Amity School of Applied Sciences, Amity University Madhya Pradesh, Gwalior, India, E-mail: pmishra@gwa.amity.edu

ABSTRACT

In the last three decades, scientists have been showing their increasing interest in the basic electrical properties of large energy and bang gap materials. Most of the literature is available on electrical behavior, charge storage and transport or most specifically the manifestation of the "electrets state" by some organic solids, waxes, insulating dielectrics, glasses and semicrystalline and amorphous polymers. Electrical properties of a polymer are partly dependent on their physical as well as chemical structure. The injected charges are trapped at different trapping sites leading to a space charge which fundamentally influences all the transport phenomena.

12.1 INTRODUCTION

Almost all polymers are amorphous or partially crystalline macromolecular organic compounds. The crystalline regions in them are microscopic (≈1000 Å) in size. Since it is not possible to obtain crystals large enough to allow proper handling, these compounds have to be studied either in the form of compressed pellets or slab cast from the melt or the thin polymer films. The increasing demand for microminiaturization of components for electronic applications has further stressed the need for growth and development of thin polymer films [1–5].

In the present chapter, experimental details are discussed right from the formation of thin films, measurement of thickness, different types of electrodes, vacuum coating of the films, various types of electrical contacts, sample holder, arrangement for current measurement, and equipment used.

12.2 THIN FILMS

Thin films are thin material layers film ranging from fractions of a nanometer to several micrometers in thickness, which can be influenced by deposition parameters in all the physical vapor deposition technique (PVD). The properties of thin films of a given material depend on the microstructure of the films, which depends on deposition parameters and the substrate material [6–8]. Depending on the interaction energies of substrate and film atoms, any of the following three growth modes can occur, as shown in Figures 12.1a, b, and c.

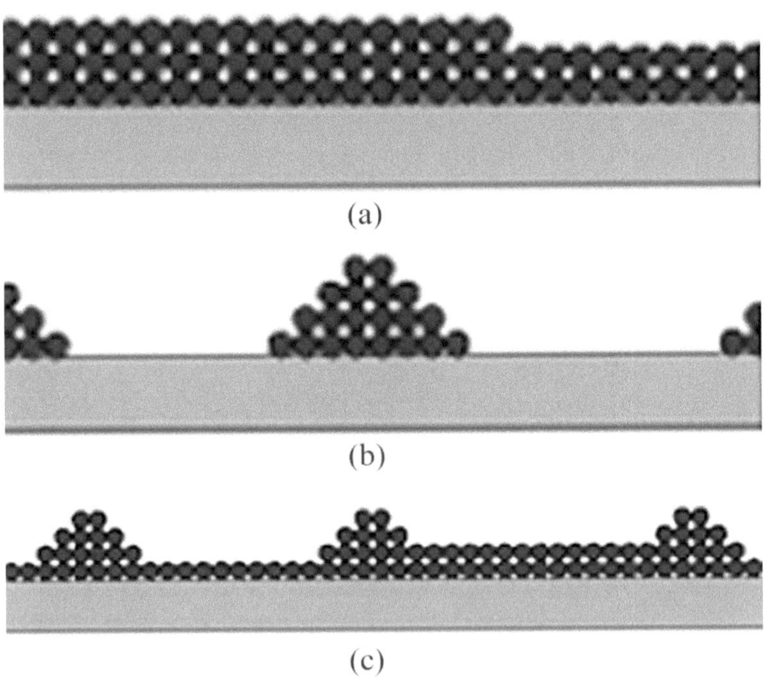

FIGURE 12.1 (a) Frank-Vander Merwe (Layer growth; ideal epitaxy); (b) Volmer-Weber: (island growth); (c) Stranski-Krastanov: (layer + islands).

a. **Layer by Layer Mode or Frank-Vander Merwe Mode:** In this two-dimensional mode, the interaction between the substrate and the layer atoms is stronger than the interaction between neighboring atoms. Each new layer starts to grow on top of another layer, only when the last one is completed.
b. **Island Growth Mode or Volmer-Weber Mode:** In this mode, the interaction between neighboring atoms is stronger than the overlayer-substrate interaction, the particles would rather form separate aggregates over the surface that grow in size and eventually coalesce (three-dimensional islands) during film growth.
c. **Layer-Plus-Island Growth Mode or Stranski-Krastanov Mode:** In this mode, the film starts to grow layer by layer in the first stage, followed by the formation of the island agglomerates occurs.

Growth modes can be systematically classified in terms of surface energies with Young's equation (Figure 12.2).

$$\gamma_S = \gamma_{FS} + \gamma_F \cos\theta \tag{12.1}$$

where, θ is the wetting angle of a liquid nucleus on a substrate, γ_S is the surface energy of substrate, γ_F is the surface energy of thin-film material, and γ_{FS} is the interfacial energy of film-substrate. The three modes of film growth can be distinguished on the basis of equation 12.1.

For layer growth, $\theta = 0$ and therefore,

$$\gamma_S = \gamma_{FS} + \gamma_F \tag{12.2}$$

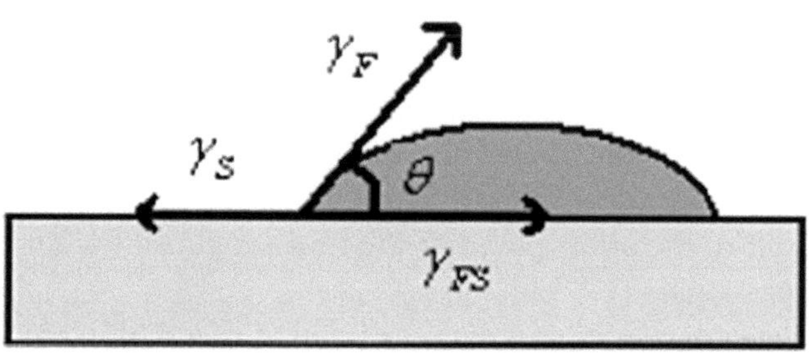

FIGURE 12.2 Schematic diagram of the surface energy of substrate (γ_S), thin-film material (γ_F), and interface energy of film-substrate (γ_{FS}).

For island growth, $\theta > 0$ and so,

$$\gamma_S < \gamma_{FS} + \gamma_F \qquad (12.3)$$

Lastly, for layer-plus-island growth,

$$\gamma_S > \gamma_{FS} + \gamma_F \qquad (12.4)$$

It occurs because the interface energy increases with film thickness.

12.2.1 SPUTTERING

Sputter deposition is one of the PVD processes for depositing thin films by sputtering, i.e., removal of atoms from a "target" (source), which then deposits onto a substrate. In other words, it is the ejection of atoms from the surface of the material (the target) by bombardment with energetic particles, the process called 'sputtering.' The ejected or sputtered atoms get condensed on a substrate and form thin films [9–11].

Sputtering process begins when inert gas atoms gets ionized due to the applied negative DC potential to the target material and these positive ions of inert gas hits the target atoms, the latter gains part of the momentum and transfer it to other atoms through further collisions, leading to a cascade which results in some of the target atoms to 'sputter' out of the target with secondary electrons. The sputtered atoms, those ejected into the gas phase, are not in their equilibrium state, therefore, they tend to condense back into the solid phase upon colliding with any surface in the sputtering chamber with maximum deposition taking place on the substrate, which is in the line of sight of the target to form a thin film. It subtends the maximum area perpendicular to the momentum of ejected target atoms and clusters. These secondary electrons collide and ionize the inert gas atoms, and plasma is generated. The initial positive ions needed to trigger the generation of secondary electrons are thought to be either the stray ions always present in the atmosphere or the ions produced by field ionization of the inert gas atoms [12–16].

Other processes associated with the bombardment of a target by highly energetic ions include:

a. Generation of secondary electrons;
b. Ion reflection at the target surface;
c. Ion implantation with the ion permanently buried into the target surface;
d. Radiation damage in the structural rearrangement varying from simple vacancies and interstitial to gross lattice defects; and
e. Emission of X-rays and photons.

These processes can be summarized, as illustrated in Figure 12.3.

To use sputtering as a useful thin film deposition process, some criteria have to be met. First, ions of sufficient energy must be created and directed towards the surface of a target to eject atoms from the surface of the material. To achieve this, an argon gas of ionization energy 15.76 eV, for example, can be used in a chamber and by application of a sufficiently large voltage between the target and the substrate; a glow discharge is set up in a way to accelerate the positive ions towards the target to cause sputtering. Secondly, the ejected materials must be able to get to the substrate with little impedance to their movement. The pressure determines the mean free path of the sputtered particles, which according to the Paschen's relation is proportional to *1/P*.

In addition to pressure, the target-substrate distance determines the scattering of the sputtered particles on their way to the substrate and also the amount of energy with which they deposit on the substrate. Generally, the average energy of the sputtered atom is 10–40 eV.

Sputtering is characterized by the sputter yield *S*, which is the ratio of the ejected atoms to the number of incoming energetic particles, which are predominantly ions [17–20]. Sputter yield depends on the energy and direction of the incident (bombarding) ions, masses of the ions and target atoms, and the binding energy of atoms in the solid. By Sigmund's theory, the sputter yield is given as

For $E < 1$ KeV,

$$S = \frac{3a}{4\pi^2}\frac{4M_1M_2}{(M_1+M_2)^2}\frac{E}{E_B} \qquad (5)$$

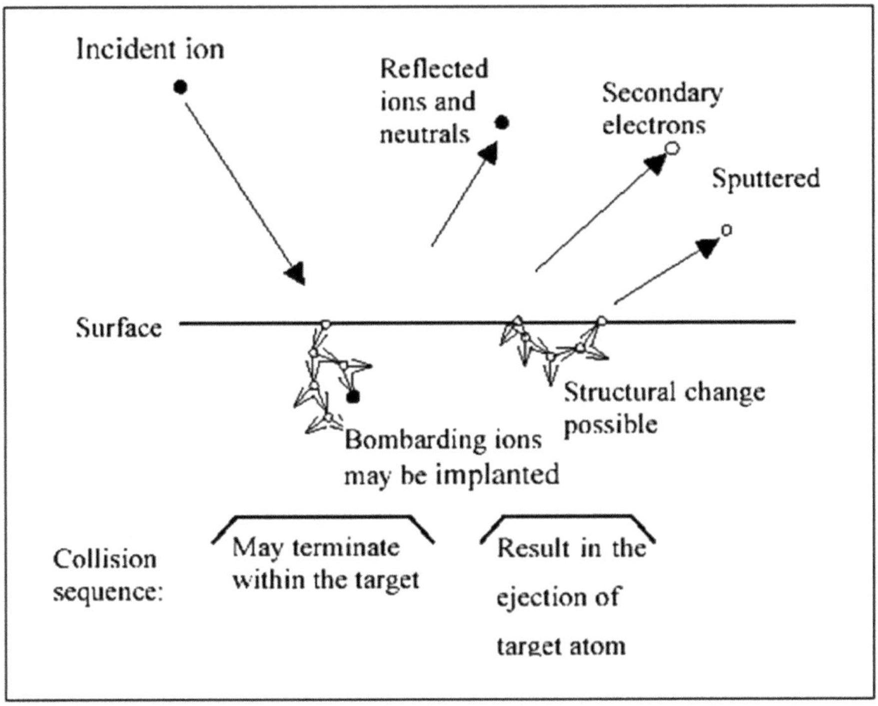

FIGURE 12.3 Processes generated by the impact of a highly energetic particle on a target surface. The collision may terminate at the target or cause particle sputtering.

Where, E_B is the surface binding energy of the target atom being sputtered, E is the ion bombardment energy, α is a measure of the efficiency of momentum transfer in collisions, and M_1 and M_2 are the masses of the positive ion of the gas and target material, respectively.

But for $E > 1$ KeV,

$$S = 3.56a \frac{Z_1 Z_2}{Z_1^{\frac{2}{3}} + Z_2^{\frac{3}{2}}} \left(\frac{M_1}{M_1 + M_2} \right) \frac{S_n(E)}{E_B} \tag{12.6}$$

where, Z_1 and Z_2 are the atomic numbers of the incident ion and sputtered target atom respectively, and S_n (E) is a measure of the energy loss per unit length due to nuclear collisions, and also it is a function of the energy as well as masses and atomic numbers of the atoms involved.

12.3 SYNTHESIS OF THIN FILMS

The choice of a deposition technique to deposit thin films usually depends on the specific characteristics of film required for a study or application of interest. Also, it is necessary to get highly uniform, nanocrystalline thin films with a mean grain size distribution. PVD techniques such as thermal evaporation, laser ablation, and magnetron sputtering are found to be very effective for depositing thin films with the aforementioned microstructural characteristics. In PVD, the synthesis of thin films is usually carried out from the same material whose nanoparticles (NPs) are to be synthesized; its purity is decided by purity of the starting materials, base vacuum, and purity of the ambient gas atmosphere. In contrast, in chemical vapor deposition (CVD) technique, some un-reacted chemicals and products other than the desired ones are often left behind with the NPs [21–30].

12.4 FILM FORMATION

Techniques of polymer film formation greatly affect their physical properties. Polymers are complex organic compounds made up of long-chain molecules that are capable of acquiring different configurations under different conditions of film formation. The ideal film is the one that possesses well-defined characteristics and reproducible physical and electrical properties [3, 4]. There are a number of techniques available; the important ones are given below:

a. Glow discharge polymerization;
b. Polymerization by electron bombardment;
c. Photolytic polymerization using ultraviolet radiation;
d. Vacuum evaporation;
e. Films from a polymer solution.

12.4.1 GLOW DISCHARGE POLYMERIZATION

At low pressure, the organic monomer vapor is subjected to continuous discharge of suitable radiofrequency. The polymerization occurs from monomer gas molecules, and the film is deposited either on the electrode or on the substrate kept near the electrode [5, 6].

Under relatively high pressure and low current density, films formed are clear, transparent, soft, and often tacky showing incomplete polymerization. However, films formed with high current density and low pressure are mechanically hard and often discolored yellow-brown, but pinholes are not present in the films formed by this technique [7–9].

12.4.2 POLYMERIZATION BY ELECTRON BOMBARDMENT

In this technique, an electron beam obtained from an electron gun is made to sweep the tile surface of a substrate exposed to the monomer vapor at a pressure of 10^{-3} torr. A continuous flow of monomer vapor is maintained through the reaction zone, and the film is deposited on the exposed surface of the substrate. The film obtained by this method is also free from pinholes [10–12].

12.4.3 PHOTOLYTIC POLYMERIZATION USING ULTRAVIOLET RADIATION

Ultraviolet rays are focused through a quartz window on a substrate at a controlled temperature. The substrate is kept in a bell jar, which can be evacuated to a pressure of about 10^{-6} torrs. Photolysis takes place when the monomer is made to flow in the reaction zone at a partial pressure of 2 Torr. The films so formed are simpler in structure than those by glow discharge polymerization [13–15].

12.4.4 VACUUM EVAPORATION

Hogarth et al. [16, 17] prepared polypropylene thin films by vacuum evaporation technique. Polypropylene was evaporated at a pressure of 6 x 10^{-5} Torr from a stainless steel boat, which was maintained at a temperature of 335°C. The film was deposited on a previously deposited copper or aluminum-based electrode. The polymer films formed by this technique are generally not free from pinholes.

12.4.5 FILMS FROM POLYMER SOLUTION

The techniques of film formation include polymerization for research purposes. Further, the polymer thin films formed by this technique may contain undesired impurities. Thin films formed by vacuum deposition techniques are also not moderately useful as these have pinholes. Thin Films of the doped polymer cannot be obtained by the above-mentioned methods since it is difficult to control the quality of dopant both during evaporation as well as during polymerization.

There are two main methods for the same.

a. **Isothermal Immersion Technique:** This method is very popular for thin film preparation. In this method, solution of suitable concentration is kept at a desirable temperature in accordance with the affecting parameters. When the film is formed, the substrate is slowly taken out and dried by hot air. The deposited film is then gently detached using a sharp knife edge [18, 19]. It is observed that the thickness of the film depends upon the concentration of the solution, its temperature, nature of the substrate, and the time for which the substrate is kept immersed in the solution.
b. **Casting from the Solution:** In this method, plane glass plates are used as a substrate for the deposition of polymer films. The glass plates are cleaned carefully by acid, water, and finally, in soap water. Subsequently, these are rinsed in the distilled water. The cleaned substrate was then dried up in hot air.

12.5 EFFECT OF THE SOLVENT

There are several aspects involved when a polymer solution makes a transition into a solid polymer film [20]. Tsilibotkin, Tager et al. [21] have shown in their study how polymer chains become more flexible by using a good solvent, thereby producing good dense films. Good quality optically smooth films free from pinholes, are obtained by using a solvent of high boiling point to ensure a slow rate of evaporation. The solutions were of sufficiently low concentration because high concentration gives different thicknesses at different spatial points [22]. The solution was evaporated at 40°C, because with higher temperatures, the high rate of evaporation makes, way for the formation of pinholes.

12.6 PRECONDITIONING OF THE FILMS

Before performing any experiment on the sample films, they were annealed in an ordinary atmosphere in an oven for 3 hours at 60°C. The samples were subjected to room temperature outgassing for 2 hours at a pressure of the order of 10^{-5} torr by a "Hind Hi Vacuum Pumping System" to remove the impurities such as water, monomer residues, and the solvent DMF.

12.7 PREPARATION OF FILMS

The solution caste method is used for thin film preparation. The solvent is allowed to evaporate in an oven at a fixed temperature for 24 h to yield the desired samples. This is followed by room temperature outgassing at 10^{-5} torr for a further period of 12 h to remove any residual solvent. Other processes will remain the same as isothermal technique.

12.8 MEASUREMENT OF THICKNESS OF THE FILM

The thickness of polymer thin films can be measured accurately by any of the following techniques:

12.8.1 OPTICAL TECHNIQUES

Under this technique, there are the following three methods:

a. **Ellipsometric Method:** In this, the thickness of transparent films is measured by mathematically analyzing the difference in the polymerization of light reflected from the film and the substrate [1, 18]. This is a nondestructive method.
b. **Interferometric Method:** In this, optical devices like Michelson's interferometer, Fabry-Perot etalon, or Newton's rings are used to measure the shift in the interference pattern between the film surface and the substrate. These methods demand the optical smoothness of the film [19].
c. **Light Sectioning Method:** In this method, light is projected at an angle of 45° from a slit over an optically smooth film formed on a substrate. If the film is opaque, then a step is made between the film and the substrate. For transparent films, the slit image is formed from the top and bottom surfaces of the film. An observing

microscope with a micrometer eyepiece placed at an angle of 450 to the sample is used for observing the separation of the slits [23, 24]. This method can be utilized for measuring the film of more than one mm thickness, but the method becomes inaccurate for the films of non-uniform thickness.

12.8.2 ELECTRICAL TECHNIQUE

In this method, vacuum evaporated electrodes are deposited on both the surfaces of the film specimen, and the capacitance of the film condenser so formed is measured using a sensitive L C R bridge [25–29]. If the high-frequency dielectric constant, area of the metallic electrodes, etc. are known, then the thickness of the sample can be very accurately known, provided the film is of uniform thickness with no pinholes. This method, therefore, is the most suitable for solution grown samples [30, 31]

12.8.3 MECHANICAL TECHNIQUES

Mechanical techniques consist of

a. **Stylus Method:** In this, a fine stylus is moved over a stepped surface formed by the edge of the film and the substrate. This stylus undergoes transverse vibrations at the step, which is recorded and amplified after being fed into an electronic circuit [32, 33]. This method is unsuitable for the film of non-uniform thickness and suffers from low accuracy.
b. **Weighing Method:** This method uses the relation between the thicknesses, mass, density, and area of the film. Since the mass is defined as the density multiplied by the volume and the area and mass of the film can be measured precisely using physical balance and vernier calipers. The thickness, t, of the film can be computed using the following formula

$$t = M/d.A \quad (12.7)$$

where, M = mass, d = density, and A = area of the film.

The sensitivity of the method depends upon the accuracy of the mass and area measurements. Also, it is not always possible to cut the substrate in well-defined areas [34, 35].

c **Micrometer Gauge Method:** This is the simplest method of measuring the thickness of a film. A number of observations are averaged to find out the exact thickness.

12.9 ELECTRODES

The polymer and electrode contact plays an important role in all electrical measurements [35, 36]. The shape of the energy band near dielectric and electrode interface decides which one of the three categories of contacts is at work. When energy band is completely horizontal, it is called a neutral contact; if a band bends downwards helping in the injection of carriers of the opposite sign then the electrode is an ohmic electrode; when the energy band bends upwards hindering the injection and neutralization of charges, the contact made is called a blocking contact.

The following four types of electrodes can be used for making electrical contacts with polymer films for the purpose of studying the charging and discharging phenomena.

12.9.1 PAINTED ELECTRODES

Conducting material paste can be used for painting the polymer film, which can be used as the electrodes. Such materials are graphite, silver, and epoxy paints. Such paints may react with the polymer and can damage it also. Therefore, the use of such painted electrodes is restricted to only that polymer, which does not react with these.

12.9.2 LIQUID CONTACT ELECTRODES

In this, nonmetallized surface of a unilaterally metalized film specimen is kept in contact with a liquid, such as water or ethyl alcohol, so that a thin uniform layer of liquid rests over the film surface. A potential is applied between the metallic electrode and the rear unmetallized surface of the film. A double charge layer is formed at the solid-liquid interface, and as a result of interaction between electrostatic and molecular forces; charge transfer to polymer film takes place. The electrode should be withdrawn, and liquid evaporated before removal of the voltage to ensure charge retention of the specimen surfaces. Recently, non-wetting liquid insulator contact electrodes have also been employed. Monocharge electrets have also been prepared using liquid contact electrodes obtained by filling one side metal-polymer gap with liquid and leaving others filled by air.

12.9.3 VACUUM DEPOSITED ELECTRODES

This is probably one of the best and convenient methods of depositing metallic electrodes of desired sizes and shapes. Metal can be evaporated in a vacuum on any metallic or nonmetallic substrate of the film specimen under study. No air gap exists between the evaporated electrodes and the substrate. The electrodes so obtained can be very conveniently used for measurements at low as well as high temperatures provided the melting point of electrode metal is higher than the temperature of measurements.

12.9.4 PRESSED METAL FOIL ELECTRODES

In this, the polymer film is sandwiched between two plane metallic foil electrodes of the desired shape and area. Springs are used to ensure uniform pressure throughout the film for proper contacts. However, in the case of polymers, the following precautions need to be taken:

a. Measurements at high temperatures should not be carried out at high temperatures, and the polymer is softened. Due to the pressure of the spring-loaded electrodes on the film, its thickness is reduced, and sometimes it results in the breakdown of the film.
b. Metallic surface of the foil needs to be cleaned, or else the foil electrode itself should be changed. Under ambient humid atmospheric conditions, practically all metallic electrodes form oxides except those of gold or platinum. This disturbs the experimental results by contaminating the film surface as the improper transfer of charges takes place through the contaminated surfaces.

12.10 VACUUM COATING OF POLYMER FILMS

The details of the unit and its operation for electrode deposition are described in the following subsections.

12.10.1 VACUUM COATING UNIT

Aluminum electrodes were deposited on polymer films using a Hind High Vacuum Coating Unit, Model 12A-4. The schematic line diagram of the unit is given in Figure 12.4. Main parts of the unit are described below:

a. Vacuum chamber consists of:
    i. Hemispherical glass bell jar with an L-shaped rubber gasket for air sealing;
    ii. Support for substrate or specimen to be coated;
    iii. Molybdenum boat or a tungsten filament for heating the material to be evaporated;
    iv. Hinged metallic shield for controlling the deposition rate of the material;
    v. High tension (HT) discharge unit (electrodes) for ionic bombardments;
    vi. Substrate rotator.
b. Rotary pump;
c. Diffusion pump;
d. Pirani gauge (for measuring coarse vacuum);
e. Penning gauge (for measuring high vacuum);
f Electric supplies to the chamber:
    i A low tension (LT) supply -for filament and boat;
    ii A HT supply -for glow discharge cleaning of the substrate;
    iii Variac for LT and HT supply.
g Valves:
    i Backing valve (V_1) connecting the rotary' pump with the diffusion pump;
    ii Roughing valve (V_2) connecting the rotary pump with the vacuum chamber;
    iii Baffle valve (V_3) for connecting the diffusion pump with the vacuum chamber;

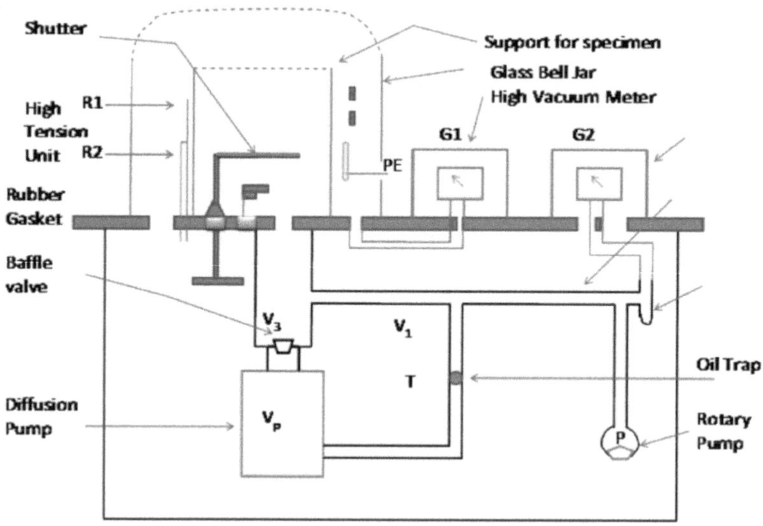

FIGURE 12.4 Vacuum coating unit.

iv Air admittance valve (V_4) for allowing air to enter the vacuum chamber;

v Gas inlet valve (V_5) for allowing the gas to enter the chamber at the desired rate.

12.10.2 OPERATION OF THE UNIT

At the start, all valves were closed, and the rotary pump was turned on. It initially evacuates the tubes connecting the rotary pump and the junction of the backing and roughing valves. When the pressure shown by the Pirani gauge attained a value less than 0.5 Torr, the backing valve V was opened for connecting the rotary pump with the oil diffusion pump circulation of water in the tubes surrounding the diffusion pump was started, and the heater of the diffusion pump was turned on.

Approximately half an hour later, with the diffusion pump ready, the backing valve V was closed, and the roughing valve V_2 was operated for connecting the vacuum chamber directly with the rotary pump. The pressure inside the vacuum chamber was allowed to fall to 10^{-2} to 10^{-3} Torr. After which the HT supply was switched on. This caused ionization of the rarefied air inside the vacuum chamber with the net result that the substrate was cleaned by ionic bombardment. The HT was switched off after 5 minutes.

The pressure inside the vacuum chamber was allowed to fall up to 0.002 Torr so as to establish a rough vacuum. After this valve, V_2 was closed, and the valve V_1 was opened again so that the rotary pump was connected to the diffusion pump for maintaining the backing vacuum. With valve V_1 open, the baffle valve V_3 was opened to connect the diffusion pump with the vacuum chamber. The vacuum was maintained by reading the pressure on the Penning gauge. When the vacuum had reached a Pressure value of 10^{-5} Torr, the LT supply was switched on. When the AL pellet kept in the spiral tungsten filament started to evaporate, the hinged metallic shield was swung out to allow the deposition of the metal vapor on the surface of the specimen. In this way, aluminum electrodes of 5.5 cm diameter were prepared.

With the deposition over, the filament and heater currents were switched off. Valve V_3 was closed, and the fine vacuum Penning gauge was switched off. After 10 minutes, the air admittance valve V_4 was opened to leak air into the vacuum chamber, making the bell jar free to be removed, and the electrodes deposited film to be taken out. The coating unit was closed down by first turning off the diffusion pump heater with the rotary pump still running and the backing valve open. After 15 minutes, when the boiler of the diffusion pump was cooled, the backing valve was closed, and the rotary pump was switched off. Finally, water circulation in the diffusion pump was stopped [31–40].

12.11 PREPARATION OF THERMOELECTRETS

The phenomenon of dielectric relaxation is due to the hindrance to the motion of dipoles and tree charges by frictional forces, and inherent inertia of motion. Therefore, on the application or removal of an electric field, a polar molecule is neither charged nor discharged immediately. Only that part of its polarization, which originates from electrons or ion displacement within the atoms or molecules, responds practically immediately. Since internal friction has been found to be exponentially proportional to the temperature, the response time to an external field is accelerated at elevated temperature and decelerated at a comparatively lower temperature. In polymers, the response time changes sharply near the glass rubber transition temperature, T, where the conformational motion of the main chain segments set in. Hence, polymers having their T_g above room temperature can be permanently charged by subjecting them to a field temperature treatment, [41–50].

Different steps for the preparation of a thermoelectric are as follows (i) the sample is heated to the desired polarizing temperature T_p; it is kept at T_p for some time (in the present case, 1 h) to reach thermal equilibrium;(ii) then, an electric field (E_p) is applied at T_p and kept on for a period of polarizing time (t_p); it is then cooled slowly under the field application to room temperature. The field is then removed. The current was recorded with an electrometer (Keithley 600 B), which was carefully shielded and grounded to avoid ground loops or extraneous electrical noise, as a function of applied field and temperature. The polarized samples were subsequently short-circuited for an arbitrary time of 10 min., so as to remove any frictional and stray charges present. The short circuit TSCs was then recorded by reheating the samples at a linear rate of 3°C/min.

The material is heated to a temperature T_p above room temperature T_r for some time until the sample reaches thermal equilibrium due to the motion of permanent dipoles and free charges. Next, at time t_0 a dc electric field of sufficient magnitude is applied, which causes an alignment of the permanent dipoles and the drift of the free charges to the electrodes [51–60].

After sometime (approximately one hour) t, the sample is cooled down to room temperature T, with the electric field still on. This causes immobilization of the main chains of the polymers, freezing in most of the permanent dipoles and charges, and the end result is a formation of a thermoelectric [61, 62]. When the electret is removed from between the electrodes, it is found to exhibit an electric field on both the surfaces which had been in touch with the electrodes. In certain polymers of low ohmic conductivity and high T_g, such as Teflon FEP, the charge such stored is retained for many years.

KEYWORDS

- **chemical vapor deposition**
- **high tension**
- **low tension**
- **nanoparticles**
- **physical vapor deposition**
- **thermoelectrets**
- **polymerization**

REFERENCES

1. Glong, R., (1970). In: Maissel, L. T., & Glong, R., (eds.), *Handbook of Thin Film, Technology.* McGraw Hi. New York.
2. Campbell, D. S., (1976). *Physics of Non-Metallic Films*. Plenum Press. London.
3. Chapman, B. N., & Anderson, J. C., (1974). *Science and Technology of Surface Coating*. Academic Press, London.
4. Harper, C. A., (1975). *Handbook of Thin Film Hybrid, Microelectronics*. McGraw Hill, New York.
5. Denaro, A. R. et al., (1968). *Europ. Polyrn. Jour., 4*, 93.
6. Williams, T., & Hayes, M. W., (1966). *Nature, 209*, 769.
7. Rose, A., J., (1964). *Appl. Phys., 35*, 2664.
8. Weash, R. C., Selby, S. M., & Hodgeman, C. D., (1964). *Handbook of Chemistry and Physics* (p. 517). Chemical Rubber Co. Pub.
9. Lyons, L. E., Kallmann, H., & Pope, J., (1957). *Chem. Soc.,* 5001.
10. Kallman. H., & Pope, M., (1960). *Nature, 186*, 31.
11. Kallman. H., & Pope, M., J., (1962). *Chem. Phy., 36*, 2482.
12. Silver, M., (1962). *Organic Semiconductors* (p. 27). McMillan Co. New York.
13. Lind, S. C., & Livigston, J., (1930). *Am. Chem., Soc., 52*. 4613.
14. White, P., (1961). *Proc. Chem. Soc.*, 327.
15. White, P., (1963). *Microelectron Reliability, 2*. 161.
16. Hogarth, C. A., & Iqbal, T., (1978). *Thin Solid Films, 51*, 45.
17. Hogarth, C. A., & Zor, M., (1975). *Thin Solid Films, 27*, 5.
18. Koutsky, J. A., Walton, A. G., & Baer, E., (1966). *J. Polymer Sci., 4*, 611.
19. Rastogi, A. C., & Chopra, K. L., (1973). *Thin Solid Films, 18*, 187.
20. Maies, K. H., & Schenermann, E. A., (1960). *Kolloid. Z*, 1171.
21. Tsilibotkina, M. V., et al., (1970). *Polymer Sci., U.S.S.R., 13*(5), 1222.
22. Frenkel, S. Y., et al., (1970). *Advances in Chemistry and Physics of Polymers.* Izd Khimiya.
23. McCrackin, F. L., (1964). *Journ. Res. NBS, 674*, 363.
24. Lakatos, A. I. J., (1970). *Appl. Phys., 46*, 1744.
25. Kondratenko, M. L., & Malik, A. L., (1974). *Instrument and Exph. Tech., U.S.A., 17*, 855.
26. Brown, R., (1960). *Am. Ceram. Soc. Bull., 45*, 206.
27. Sawa, S., et al., (1974). *Journ. Polym. Sci. Phy., 12*. 1231.
28. Rencroff, P. J., & Takahasi, K., (1975). *Journ. Non-Cryst. Solids, 1*, 71.
29. Rencroff, P. J., & Ghosh, S. K., (1974). *Journ Non-Cryst. Solids, 15*, 399.
30. Lupu, A., Giuergea, M., J., (1974). *Polym. Sci. Polym. Phys., 12*, 2399.
31. Bhatnagar, S. N., & Shrivastava, A. P., (1972). *Ind. Journ. Pure. Appl. Phys., 10*, 360.
32. Jain. V. K., Gupta, C. L., & Jain, R. K., & Indian, J., (1978). *Pure and Appl. Phys., 16*. 625.
33. Vyas, K. D., (1992). *Electrical Properties of High Polymers With Special Reference to Electret Effect in Pure and Doped PMMA*. Research thesis submitted in-1992, in Rani Durgawati Vishwavidyalaya Jabalpur. India.
34. Wagner, M. F., (1978). *Thermochemic. Acta, 23*, 93.

35. Jain, K., & Chopra, K. L., (1973). *Phys. Stat, Sol. (a), 20*, 167.
36. Gazso, J., (1974). *Thin Solid Films, 21*, 43.
37. Atkinson, P. J., & Flemming, R. J., (1980). *Journ. Phys. D. Appl. Phys., 12*, 625.
38. Pearson, J. M., (1989). *Vinylcarbazole Polymers In Encyclopedia of Polymer Science and Engineering* (p. 257). Wiley: New York.
39. Pearson, J. M., ibid.
40. Hallensleben, M. L., (1992). Other polyvinyl compounds. In: Elvers, B., Hawkins, S., & Schultz, G., (eds.), *Ullmann's Encyclopedia of Industrial Chemistry* (5th edn.). VHS: New York, A21, 743.
41. Mark, H. F., et al., (1989). *Encyclopedia Polymer Science and Engineering* (Vol. 17, p. 272). John Wiley and Sons, New York.
42. Klopffer, W., (1971). *Kunstoffe, 61*, 533.
43. Cornish, E. H., (1963). *Plastics, 28*, 61.
44. Pearson, J. M., & Stolka, M., (1981). *Polymer Monographs* (Vol. 6). Gordon and Breach, New York.
45. Jacobi, H., (1959). *Kunstoffe, 43*, 381.
46. Penwell, R., Ganugly, B., & Smith, T., (1978). *J. Polym. Sci., Macromol. Rev., 13*, 63.
47. Davidge, H. J., (1959). *Appl. Chem., 9*, 553.
48. *Data Sheet No. 263,* (1990). Poly (N-vinylcarbazole). Polysciences, Inc., Warrington, Penn.
49. *BASF Data Sheet*, (1971). PolyvinylcarbazoleĐLuvican1.
50. Winkelmann, D., Pai, D., Crooks, W., Pennington, K., Lee, F., Bräuninger, A., Brabandere, L., Verelst, J., Frass, W., Hoffmann, H., Sprinstein, K., Steppan, H., Stoudt, T., & Allen, D., (1986). Imaging technology. In: Elvers, B., Hawkins, S., & Schultz, G., (eds.), *Ullmann's Encyclopedia of Industrial Chemistry* (5th edn., Vol. A13, p. 571). VHS: New York.
51. Lardon, M., Lell-Doller, E., & Weigl, J. W., (1967). *Mol. Crystl., 2*, 241.
52. Gill, W. D., (1972). *J. Appl. Phys., 43*, 5033.
53. Weiser, G., (1972). *J. Appl. Phys., 43*, 5028.
54. Melz, P. J., (1972). *J. Chem., Phys., 57*, 1694.
55. Zhang, Y., Wada, T., & Sasabe, H., (1998). *J. Mater. Chem., 8*(4), 809.
56. *Holography*, (1993–1997). In Microsoft® Encarta® 98 CD-ROM Encyclopedia, Microsoft: Seattle.
57. Hoegl, H., (1965). *J. Phys. Chem., 69*, 755.
58. Pearson, J. M., & Stolka, M., (1981). Poly(N-vinylcarbazole). In: *Polymer Monographs No. 6*. Gordon and Breach: New York.
59. Zhang, Y., Wada, T., & Sasabe, H., (1998). *J. Mater. Chem., 8*(4), 809.
60. Zhao, C., Park, C., Prasad, P. N., Zhang, Y., Ghosal, S., & Burzynski, R., (1995). *Chem. Mater., 7*, 1237.
61. Mishra, J., & Mishra, P. K., (2015). *Advanced Science Letters, 21*(9), 2933–2936.
62. Mohan, K., & Pankaj, K. M., (2017). *Conf. Proceedings Trace* (pp. 632–635). ASET, Amity Noida.

CHAPTER 13

Pharmaceuticals Polymers in Medicine

DIVYA SINGH[1] and NARENDRA PAL SINGH CHAUHAN[2]

[1]*Department of Chemistry, Amity School of Engineering and Technology, Amity University Madhya Pradesh, Gwalior, India, E-mail: drdsingh18@gmail.com*

[2]*Department of Chemistry, Bhupal Nobel's University, Udaipur, Rajasthan, India, E-mail: narendrapalsingh14@gmail.com*

ABSTRACT

Polymer degradation gives nontoxic biproducts. These products have good mechanical strength. Therefore such biproducts are applicable in bone tissue engineering, also helps in bone growth when used in combination with bi-nanoparticles and hydrogels can also be prepared using such byproducts which help in drug delivery. Such polymers are hydrophilic and semi-permeable in nature.

13.1 INTRODUCTION

The core properties of specific polymers like branching, networking, the property of 3D structure formation by multiple assemblies of simple structural units are the most acceptable class of polymer in the medicinal and biotechnology field. The classical natural polymers such as cellulose, keratin of hairs, etc. are used in medicine since past decades [1]. Synthetic polymers procured attention as well as allured its usefulness in the area of medicine worldwide. Various homo and copolymers are synthesized at a different concentration, which fulfills the mechanical, structural, and biological criteria so that the considerable utility of such material can be extracted for future development in the field of science [2]. The advancement in mechanical properties of such polymers increases its use in drug delivery systems and devices, in preparation of tubes used in vascular tracts, various devices used in medicine like surgical equipments, blood clot removal devices. The oriented fiber of the same material is integrated into the matrix, and this gives matrix the mechanical strength and self-reinforcement [3]. The advanced mechanical properties like the shape of polymers can be deformed and then reformed back to original shape upon special stimulus like pH, temperature, magnetic field, or light. Such polymers have various bio-medicinal applications like drug delivery as well as bio-medicinal equipments like drug carrier's devices, vascular stents, sutures, clot removal equipments, for ducts, occlusion device, and in various orthodontic materials [4]. The synthetic and natural pharmaceutical biopolymers have embraced to a greater level in recent years.

Various functionalized polymers with specific characteristics are also found to be useful in medicinal biotechnology. The bioactive polymers such as hexacellulose, cellulose, dextrin, etc. are used in biomembrane synthesis, which can act as semi-permeable membrane and can be useful in dialysis and drug delivery. The membrane property can be enhanced like swelling, collapsing of pores of the membrane in response to various experimental conditions like pH and temperature. This helps in a better understanding of proper drug release through the membrane. Such functionality of polymer can benefit medicine to a great extent to understand the concept of drug delivery to target sites as well as enlightening it's the bio-technical applications [5]. Apart from such properties, there are various other properties of polymers that are directly and indirectly adding up to the growth of its use in biotechnology and medicine. We cannot separate pharmaceutical and medicine from biotechnology as the methodology used in biotechnology for the synthesis of such functionalized polymers helps in enhancing the properties of these materials in medicinal as well as bio-medicinal fields [5]. Likewise, its importance as bio-tissue engineering material due to its carbon-based properties of synthetic polymers, nanohybrid structures, etc., is gaining much attention nowadays. Due to its greater interaction with the targeted body, it had caused some interference during the work process due to the formation of degradation by-products, or additives formation during biochemical pathways. The active substituent's in polymer makes the polymer bio-functional. The upcoming plasma technologies develop surface bio functionality, which modifies surface phenomena for greater biocompatibilities and optimization or cumulation of the automated and mechanical properties of the bulk. Thus functionalized polymers have vast bio-medicinal applications. Some synthetic biodegradable polymers used in this area have idealistic characterizes as they stay in the body for the long-time period and severe their complete functionality, less toxic, and can be degraded internally and disappear without any surgical methods [6, 7]. The applications of biodegradable polymers in ligament accession or amplification and use by orthopedic in the past was the prime inducement for the use of biodegradable polymers [8]. The vascular stents and use of polymers in drug delivery as microcarriers and its faster degradation increases its application at micro as well as macroscopic level since last decades [9–11]. In the current pharmaceutical field, the most important application of such unique polymers is for controlled drug release and bio-implants, which is discussed in the chapter.

13.2 CLASSIFICATION OF BIO-MEDICINAL PHARMACEUTICAL POLYMERS

Polymer-based bio-medicinal pharmaceutical products are in great demand, and its wide range of applications in the various medicinal zone like coating, taste-making for oral dosages, packaging, fabrication in drug delivery devices, etc. is highly embraced. Thus such biomedical polymers are classified in some major classes as discussed below.

Three basic classifications of different types of bio-medicinal pharmaceutical polymers are categorized, which are as follows:

- Based on the applicability of pharmaceutical polymers;
- Based on the solubility of pharmaceutical polymers;
- Based on colloidal and matrix-based properties.

13.2.1 BASED ON APPLICABILITY OF PHARMACEUTICAL POLYMERS

These categories of polymers are further classified on the basis of various application and usability, which are as follows:

- Polymers used conventionally as a standard dosage;
- Polymers used as controlled release dosage form;
- Polymers in drug packaging.

13.2.2 POLYMERS USED CONVENTIONALLY AS STANDARD DOSAGE

Though controlled drug release is an advanced method for dosage intake, but still conventional method is an easy and usable method for oral intake of medicine. Oral dosage forms like tablets, capsules, and liquids are still used at wider aspects due to its low cost, which is categorized in the conventional methods for drug intake. Such oral dosages require these classes of polymers for preparation for easy release into the body. Tablets are used as an oral drug commonly. Its formation requires the incorporation of polymeric material. Apart from solid dosage forms, liquid dosage forms are also commonly used. In liquid dosage form its acts as rheology modifiers. It controls the viscosity of aqueous drugs, and it also helps in stabilizing suspension mediums as well as in the preparatory step of granulation of solid doses [12].

13.2.3 POLYMERS USED AS CONTROLLED RELEASE DOSAGE FORM

In controlled drug release method, the drug absorption (by diffusion, dissolution, osmosis, and ion-exchange) and drug concentration in blood can be altered, thus such polymers are used which act as modifier in drugs to alter release rates and check changes in drug plasma concentration and how it is affecting patient and any side effects or any increased compliance. It consists of an active agent and polymer matrix which regulates drug release. Such polymers are modified and tailored accordingly on the basis of properties like charge density, ion specificity, molecular size, hydrophobic behavior, specified functionalized groups, biocompatibility, and degradability. Thus reduces many bio medicinal demerits and boosts pharmacological aspects of the drugs. The polymers used in controlled release dosage of protein drugs have gained much attention. Since such drugs are bioactive. The chronic diseases can be healed by such drugs totally depend upon the novel delivery system for routine dose administration other than injection. Thus such polymer controls drug transportation.

13.2.4 POLYMERS IN DRUG PACKAGING

This category of polymers acts as a major functionalized unit for drug packaging in pharmaceutical and bio-medicinal products. Permeability for gas, flexibility, and transparency are concerns properties in packages. They are adaptable to reform its shape. The thin, flexible films are usually produced from cellulose derivatives, polyvinyl chloride, polyethylene, polypropylene, polyamide (nylon), polystyrene, polyesters, polycarbonate, polyvinylidene chloride, polyurethanes, etc. Such polymer materials forms are generally heat resistive and heat sealable and can laminate other materials which can tightly pack the goods by proper wrapping followed by heat treatments. Rigid and tough packaging materials such as bottles, trays, cups, and various closures have sufficient strength and inflexibility.

13.2.5 BASED ON SOLUBILITY OF PHARMACEUTICAL POLYMERS

Some are water-soluble, and some are water-insoluble, which develops major properties in polymers so that a broader research relating its application can be studied.

They are of following types:

- Water-soluble polymers;
- Water-insoluble biodegradable polymers;
- Starch-based polymer;
- Plastics and rubbers.

13.2.6 WATER-SOLUBLE POLYMERS

Water-soluble polymers like, for example, poly(ethylene oxide) act as a coagulant, flocculent, swelling agents in medium, polyvinyl pyrrolidone, polyvinyl alcohol can be used as binder, in coating, tablet formation, water-soluble packaging, granule formation, in plasma replacement, poly-isopropyl poly-acrylamide, poly-cyclopropyl methacrylamide and polyethylene glycol (PEG) can be used as gel formation agent, plasticizer etc. All these polymers have pharmaceutical and medicinal importance.

13.2.7 WATER-INSOLUBLE BIODEGRADABLE POLYMERS

Water-insoluble polymers like, for example, lactide-co-glycolide polymers for protein delivery, which are physically strong and highly bio-compatible and functions are protein carriers or drug delivery vehicles and also as DNA and RNA carriers. They are also used in nanoparticle forms. They are nontoxic. Such polymers are water-insoluble, so they are converted to carbon dioxide and water under specific experimental conditions. They undergo bio-degradation, bio-fragmentation, assimilation, and bio-deterioration. They are more focused to degrade in soil or compost than following conventional methods.

13.2.8 STARCH-BASED POLYMER

Sodium starch glycolate used as a super-disintegrant for tablets and capsules in oral delivery, starch glidant as diluents in tablets and capsules, as a tablet binder. Starch-based biodegradable polymers are also called as medical polymeric materials and have many advantages like good biocompatibility, biodegradability. Degradation of polymer gives nontoxic products with good mechanical properties is applicable in bone tissue engineering (TE) to help in bone growth when, in combination with bio nanoparticles (NPs), hydrogels are prepared, which help in drug delivery—such polymers are hydrophilic and semi-permeable.

13.2.9 PLASTICS AND RUBBERS

Some relevant application in this category of polymer is explained below through some examples like Poly-cyano acrylate act as biodegradable adhesives for tissue cells in surgery, a drug carrier in nano- and microparticles. Poly-chloroprene can be used to form septum or membrane for injection, plungers for syringes, and components for valves. Poly-isobutylene pressure-sensitive adhesives for transdermal delivery, silicones pacifier, therapeutic devices, implants, medical-grade adhesive for transdermal delivery, polystyrene used in Petri-dishes and containers used for cell culture. Poly-(methyl methacrylate) for hard contact lenses formation, Poly-hydroxy ethyl methacrylate uses in soft contact lenses, and polyvinyl chloride used in blood bags, hoses, and tubing formations. Thus, it is concluded that there are multiple uses such as materials in medicine, medicinal equipments preparation, and pharmaceutical benefits for mankind.

13.2.10 BASED ON COLLOIDAL AND MATRIX-BASED PROPERTIES

The following class of polymers includes polymer which forms colloids and matrix which are useful in cosmetics and controlled drug delivery, can be used as mucoadhesive dosage forms, rapid release dosage, oral, and topical pharmaceutical products, coating, and capsule formation are as follows:

- Hydrocolloids (e.g., carrageenan, chitosan, pectinic acid, and alginic acid, etc.).
- Cellulose-Based Polymers (e.g., hydroxypropyl methylcellulose, hydroxyethyl, and hydroxypropyl cellulose, etc.).

13.3 APPLICATION AS BIOMEDICAL IMPLANT

Bio-implants upgrade the nature of our mankind by expanding the usefulness of basic frameworks for life expectancies. Over the restorative business, different bio-medicinal implants and equipments have been considered and created for saving a life. The man-made articles, for example, knee inserts, engineered veins, etc., enhance the usefulness of human organs; for example, the pacemaker for heart, the focal objective of these gadgets is focused towards the conservation of life. These applications likewise shift regarding their situation and positions inside the body. A considerable lot of these gadgets are set in areas of high mechanical pressure, for example, in the joints amid bone substitution or in districts of high synthetic and electrical movement, for example, the utilization of neuro-prosthetics. Arrangement of each embed or gadget brings has an alternate arrangement of necessities in the structure and material choices. As per the United States Food and Drug Administration, a therapeutic gadget is an instrument, device, execute, machine, embed, in vitro reagent, or other comparable or related article which is used in the determination or treatment or counteractive action of any kind of infection.

The US Food and Drug Administration (FDA) isolates the devices with conceivable manufactured polymers are classified in Table 13.1.

TABLE 13.1 Some Common Medical Bio-Implants

Category	Biomedical Device	Synthetic Polymer in Devices
Anesthetic	Catheters	Poly-amide, Poly-ethene, Poly-tetrafluoroethylene
Cardiovascular	Pacemaker for heart, Catheters, implantable cardioverter/ defibrillator, artificial heart valves, and blood vessels	Poly-ethene, Polypropylene, Poly-tetrafluoroethylene, Polyamide, Poly-ethylene terephthalate, Polydimethylsiloxane, Polyhydroxyalkanoates
Dental	Implants for dental purpose	Polymethylmethacrylate
Ear, nose, and throat	Cochlear implants related to cochlear and stapes, Nasal track	Polydimethylsiloxane, Silicone, Poly(p-xylylene), Polyethene
Urinary and Gastroenterological purpose	Hernia or vaginal mesh, Penile and artificial urinary sphincter implants, Neuro-stimulator in sacral nerve, stimulation	Polydimethylsiloxane, Polyethene, Polytetrafluoroethene, Polyamide, Polyhydroxyalkanoates, Silicone rubber, Polypropylene
Surgery for general and plastic	Blood vessels by synthetic method, breast implants, cheek, jaw, lip, and chin implants, hip, and titanium surgical bioimplant	Polypropylene, Polyethylene terephthalate, Polytetrafluoroethene, Silicone rubber, Polydimethylsiloxane
Hematological and pathological	Central catheter, Central venous access device	Polyethene, Polytetrafluoroethene, Polyamide
Neurological purpose	Catheters, Generation of pulse by stimulation of brain and brain nerves, Neuro-prosthetics	Polyimides, Polydimethylsiloxane, Poly(p-xylylene), Liquid crystal polymers, Polyethene, Polytetrafluoroethene, Polyamide, Polyhydroxyalkanoates
Obstetric and gynecologic	Intravaginal, intra-uterus rings, contraceptive implant, surgical mesh bioimplants, fetal micro-pacemaker	Silicone rubber, Polyurethane, Polypropylene
Ophthalmic	Intra vitreal implant, Retina correction devices, Artificial intraocular lens, ophthalmic implant, catheters	Polymethylmethacrylate, Polyethylene, Polytetrafluoroethylene, Polyamide
Orthopedic	Orthopedic bioimplants	Poly-ethene, Polyether ether ketone, Poly-hydroxy alkanoates

Smart materials will be materials intended to be receptive to outer improvements, for example, pH, temperature, electric or attractive fields, organic signs, and neurotic anomalies. In light of ecological boosts, at least one of their physicochemical, as well as organic qualities are changed [13] Some models include: shape-memory materials that can recuperate their unique structure, pH-responsive polymeric materials that cause mark change dissolvability, water absorption ratio, furthermore, surface charge because of ecological pH change, and immobilized bioactive atoms on materials that can upgrade their natural reaction to the miniaturized scale condition. These bioactive polymers are also found to be antimicrobial in nature, which adds up to its excellence in pharmaceutical industries and biomedicine. A various number of recently used such polymers like 4-(1-nitrosoethylidene)-1H-pyridine, P-chloroacetophenone oxime, 8-hydroxyquinoline, furfural, and other substituted acetophenones, etc. have been reported [14–18].

The appropriateness of the physical properties of polyurethane can prompt their selection in biomedical applications. Moreover, the physicochemical qualities of polyurethane can be effortlessly tuned through adjustments of the synthetic parts, because of the huge similarity of polyurethane combination with a wide assortment of useful monomers. Because of these two highlights, different brilliant polyurethanes have been produced for biomedical applications that can capacity to be thermo-responsive, have shape memory, and be pH-responsive, self-healing, self-repairing, bioresponsive, and self-cleaning [19–30]. Various self-healing polyurethane materials have been developed for biomedical purposes. Poly (methyl methacrylate)-co-[poly(methyl methacrylate)-graft-(oligo-caprolactone)] urethane systems containing a Diels-Birch adduct unit was synthesized [31]. The systems show reversible scratch-recuperating properties by the crosslinked (at Diels-Birch apportion temperature, 70°C) and de-crosslinked (at retro-Diels-Birch response temperature, 130°C) structure of the Diels-Birch adduct unit which is characterized by FT-infrared spectra which has shown that the maleimide twofold bond crest at 654 cm^{-1} was seen after de-crosslinking at 130°C and vanished again in the wake of crosslinking at 70°C.

Some substituted alkyl-based polymers, for example, N-isopropyl acrylamide and N-(hydroxymethyl) acrylamide, helps in shrewd stent formation contains nanofibrous units intact [32].

Perera and colleagues [33] reported an attractively incited medication conveyance framework, in view of PVA polyvinyl alcohol—MNP citrus extract covered Fe_3O_4 attractive NPs strands, created by means of mixture gyration. The strands are made of biocompatible segments with the end goal to target biomedical applications, particularly, tranquilize discharge. The capacity of these filaments to be impelled by means of an attractive outside field is illustrated, alongside a broad portrayal of its physico-compound properties. The arrival of controlled amounts of acetaminophen, by means of attractive incitation of this material, is investigated (Figure 13.1).

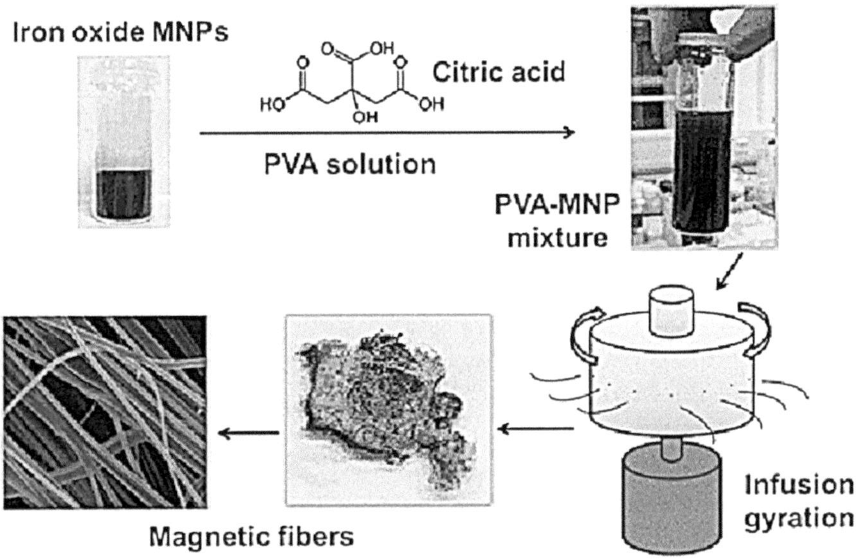

FIGURE 13.1 PVA-MNP fibers for drug delivery applications.

(Source: From Perera et al. [33] © 2018, with kind permission of ACS applied materials and interfaces; https://pubs.acs.org/doi/abs/10.1021%2Facsami.8b04774).

Polymer micelles are a standout amongst the most explored nanocarriers for medication conveyance; many have entered clinical preliminaries, and some are in facility utilize. However, their conveyance frameworks have not yet demonstrated the normal high restorative adequacy in centers. Sun and associates the clinically applicable polyethylene glycol-square poly(ε-caprolactone) and PEG-square poly(DLlactide) micelles for enhanced therapeutics for enhancing micelle solidness in the blood is essential for surface functionalizations, for example, presenting focusing on ligands [34].

13.4 SUMMARY AND CONCLUSION

Polymeric substances are not only used as major ingredients in medicine and biotechnology in form of drug dosages but also in packaging materials and processing aids for drugs and capsule. The broader use of polymers as in coating, packaging, encapsulation, as dosage forms, etc., increased its wide spread used worldwide. The conventional dosage forms of polymeric drugs contain polymer coatings or covering and use as natural, biodegradable polymers with wide range application as listed in given chapter, explains its great history in pharmaceutical market. The use of polyethylene, polypropylene, and poly vinyl chloride is occurring science past in pharmaceutical packaging, polyethylene. Then, the use of more environmental, bio-friendly, biodegradable polymers has caught fire to replace above mentioned polymers. The broad acceptance of polymers such as natural, synthetic, biodegradable polymers, etc., in numerous fields in pharmaceutical industry for the medicinal production of useful materials focused attention of its use in controlled drug delivery system or dosages release in the targeted sites under study. The incorporation of polymer science in pharmaceutics uplifted the growth in controlled drug delivery so that many ways can be develop to fight against chronic diseases and can find cure in that regard. This increases sophistication and clearance for the development of novel and advanced technology in multiple fields. This has achieved development in self-regulated drug delivery, long-term delivery of protein drugs, and drug targeting to specific organs in the body. The latest technology includes bio-implants which also include polymers in broader aspect. These help in making survival and life expectancies easier. This also increases the economical growth in bio-medicinal field as well as developed possibilities of replacing organs in body to conserve human life using this latest technology which is not possible without the use of polymers. This requires advanced well designed system which need polymeric vehicle for growth and novel research. Polymers have become the integral unit or backbone for bringing the change in era of pharmaceutical industries and medicine.

Polymer family covers the widest range of application in field of biotechnology especially opening gate way in area of medicine. The nano, hybrid biodegradable polymeric materials are of great importance in field of medicinal technology. This reduces the use of traditional materials such as glass metals, etc. The polymers such as natural, synthetic, hybrid, polymeric nano forms, biodegradable, etc., are used. They are modified and are tuned use with different physical, chemical and biological properties for the advancement in biotechnology. The alkyl based polymer like butyryl-trihexyl-citrate and Didi(2-ethylhexyl)phthalate a used as alternative plasticizer of polyvinyl chloride in blood bags. Expanded polymeric tetra flouro ethyelene is used for vascular grafts, surgical meshes, ligament, and tendon repair. Polyether polyether ether ketone is useful for hard stable polymer for orthopedic applications or inner lining of catheters. Poly-methyl methacrylate is used as hard methacrylate as bone cement, lenses and for membranes for kidney dialysis. The synthetics polymers having excellent biological, physical and chemical properties with low toxicity are broadly used in biotechnology applications. The synthetic polymers are also spreading its end use in various with specific applications such as blood bank bags preparation, in surgical equipments preparation, mechanical valve preparation, in ligament and tendon repair or substitution, etc. Various materials prepared using thermal stable polymers like catheters bone cement, eye lenses, membrane used in dialysis, etc., explains the polymer needfulness in biotechnology. Polymeric tools are used for controlling drug release rate. Such polymers are used as masking agents, stabilizer, and protective agents in oral dosages for drug delivery. It act as binder for solid dosage and it can also affect the viscous nature of liquid dosages which can change the flow properties of liquid dosage. This uniqueness of such macromolecules increases its ranking as pharmaceutical polymers and it become highly applicable in biomedicine. This functionalized macromolecule becomes the foundation for greater understanding for drug type, drug design and drug delivery, etc. This chapter shed light on the use of polymers in medicine and biotechnology.

KEYWORDS

- **drug packaging**
- **Food and Drug Administration**
- **nanoparticles**
- **pharmaceutical polymers**
- **polyethylene glycol**
- **tissue engineering**

REFERENCES

1. Hazer, D. B., Kılıçay, E., & Hazer, B., (2012). Poly (3-hydroxyalkanoate) s: Diversification and biomedical applications: A state of the art review. *Materials Science and Engineering: C, 32*(4), 637–647.
2. Ma, Y., Wang, R., Cheng, X., Liu, Z., & Zhang, Y., (2013). The behavior of new hydrophilic composite bone cements for immediate loading of dental implant. *Journal of Wuhan University of Technology-Mater. Sci. Ed., 28*(3), 627–633.
3. Kim, J. H., Min, K. S., Jeong, J. S., & Kim, S. J., (2011). Challenges for the future neuroprosthetic implants. In: 5th *European Conference of the International Federation for Medical and Biological Engineering* (pp. 1214–1216). Springer, Berlin, Heidelberg.
4. Luo, C., Cao, G. Z., & Shen, I. Y., (2013). Development of a lead-zirconate-titanate (PZT) thin-film microactuator probe for intracochlear applications. *Sensors and Actuators A: Physical, 201*, 1–9.
5. Zhou, J., Huang, X., Zheng, D., Li, H., Herrler, T., & Li, Q., (2014). Oriental nose elongation using an L-shaped polyethylene sheet implant for combined septal spreading and extension. *Aesthetic Plastic Surgery, 38*(2), 295–302.
6. Moalli, P., Brown, B., Reitman, M. T., & Nager, C. W., (2014). Polypropylene mesh: Evidence for lack of carcinogenicity. *International Urogynecology Journal, 25*(5), 573–576.
7. Lee, S. W., Min, K. S., Jeong, J., Kim, J., & Kim, S. J., (2011). Monolithic encapsulation of implantable neuroprosthetic devices using liquid crystal polymers. *IEEE Transactions on Biomedical Engineering, 58*(8), 2255–2263.
8. Hassler, C., Boretius, T., & Stieglitz, T., (2011). Polymers for neural implants. *Journal of Polymer Science Part B: Polymer Physics, 49*(1), 18–33.
9. Rahimi, A., & Mashak, A., (2013). Review on rubbers in medicine: Natural, silicone and polyurethane rubbers. *Plastics, Rubber and Composites, 42*(6), 223–230.
10. Malcolm, R. K., Edwards, K. L., Kiser, P., Romano, J., & Smith, T. J., (2010). Advances in microbicide vaginal rings. *Antiviral Research, 88*, S30–S39.

11. Kaur, M., Gupta, K. M., Poursaid, A. E., Karra, P., Mahalingam, A., Aliyar, H. A., & Kiser, P. F., (2011). Engineering a degradable polyurethane intravaginal ring for sustained delivery of dapivirine. *Drug Delivery and Translational Research, 1*(3), 223.
12. Corpet, D. E., Parnaud, G., Delverdier, M., Peiffer, G., & Taché, S., (2000). Consistent and fast inhibition of colon carcinogenesis by polyethylene glycol in mice and rats given various carcinogens. *Cancer Research, 60*(12), 3160–3164.
13. Wang, F., Li, Z., Lannutti, J. L., Wagner, W. R., & Guan, J., (2009). Synthesis, characterization and surface modification of low moduli poly (ether carbonate urethane) ureas for soft tissue engineering. *Acta Biomaterialia, 5*(8), 2901–2912.
14. Chauhan, N. P. S., (2013). Structural and thermal characterization of macro-branched functional terpolymer containing 8-hydroxyquinoline moieties with enhancing biocidal properties. *Journal of Industrial and Engineering Chemistry, 19*(3), 1014–1023.
15. Chauhan, N. P., Ameta, R., & Ameta, S. C., (2011). Synthesis, characterization, and thermal degradation of p-chloroacetophenone oxime based polymers having biological activities. *Journal of Applied Polymer Science, 122*(1), 573–585.
16. Chauhan, N. P. S., (2012). Preparation and thermal investigation of renewable resource based terpolymer bearing furan rings as pendant groups. *Journal of Macromolecular Science, Part, A., 49*(8), 655–665.
17. Chauhan, N. P., & Ameta, S. C., (2011). Preparation and thermal studies of self-crosslinked terpolymer derived from 4-acetylpyridine oxime, formaldehyde and acetophenone. *Polymer Degradation and Stability, 96*(8), 1420–1429.
18. Chauhan, N. P., (2012). Isoconversional curing and degradation kinetics study of self-assembled thermo-responsive resin system bearing oxime and iminium groups. *Journal of Macromolecular Science, Part, A., 49*(9), 706–719.
19. Yu, S., He, C., Ding, J., Cheng, Y., Song, W., Zhuang, X., & Chen, X., (2013). pH and reduction dual responsive polyurethane triblock copolymers for efficient intracellular drug delivery. *Soft Matter, 9*(9), 2637–2645.
20. Sivakumar, C., & Nasar, A. S., (2009). Poly (ε-caprolactone)-based hyperbranched polyurethanes prepared via A2+ B3 approach and its shape-memory behavior. *European Polymer Journal, 45*(8), 2329–2337.
21. Shumaker, J. A., McClung, A. J. W., & Baur, J. W., (2012). Synthesis of high temperature polyaspartimide-urea based shape memory polymers. *Polymer, 53*(21), 4637–4642.
22. Singhal, P., Rodriguez, J. N., Small, W., Eagleston, S., Van De Water, J., Maitland, D. J., & Wilson, T. S., (2012). Ultra low density and highly crosslinked biocompatible shape memory polyurethane foams. *Journal of Polymer Science Part B: Polymer Physics, 50*(10), 724–737.
23. Miyazu, K., Kawahara, D., Ohtake, H., Watanabe, G., & Matsuda, T., (2010). Luminal surface design of electrospun small-diameter graft aiming at in situ capture of endothelial progenitor cell. *Journal of Biomedical Materials Research Part B: Applied Biomaterials, 94*(1), 53–63.
24. Zhang, C., Zhao, K., Hu, T., Cui, X., Brown, N., & Boland, T., (2008). Loading dependent swelling and release properties of novel biodegradable, elastic and

environmental stimuli-sensitive polyurethanes. *Journal of Controlled Release, 131*(2), 128–136.
25. Zhou, H., Xun, R., Liu, Q., Fan, H., & Liu, Y., (2014). Preparation of thermal and pH dually sensitive polyurethane membranes and their properties. *Journal of Macromolecular Science, Part, B., 53*(3), 398–411.
26. Zhao, C., Nie, S., Tang, M., & Sun, S., (2011). Polymeric pH-sensitive membranes—a review. *Progress in Polymer Science, 36*(11), 1499–1520.
27. Gu, L., Wang, X., Chen, X., Zhao, X., & Wang, F., (2011). Thermal and pH responsive high molecular weight poly (urethane-amine) with high urethane content. *Journal of Polymer Science Part A: Polymer Chemistry, 49*(24), 5162–5168.
28. Wandera, D., Wickramasinghe, S. R., & Husson, S. M., (2010). Stimuli-responsive membranes. *Journal of Membrane Science, 357*(1&2), 6–35.
29. Wu, W., Zhu, Q., Qing, F., & Han, C. C., (2008). Water repellency on a fluorine-containing polyurethane surface: Toward understanding the surface self-cleaning effect. *Langmuir, 25*(1), 17–20.
30. Zheng, J., Song, W., Huang, H., & Chen, H., (2010). Protein adsorption and cell adhesion on polyurethane/Pluronic® surface with lotus leaf-like topography. *Colloids and Surfaces B: Biointerfaces, 77*(2), 234–239.
31. Kim, S. Y., Lee, T. H., Park, Y. I., Nam, J. H., Noh, S. M., Cheong, I. W., & Kim, J. C., (2017). Influence of material properties on scratch-healing performance of polyacrylate-graft-polyurethane network that undergo thermally reversible crosslinking. *Polymer, 128*, 135–146.
32. Aguilar, L. E., Ghavami, N. A., Park, C. H., & Kim, C. S., (2017). On-demand drug release and hyperthermia therapy applications of thermoresponsive poly-(NIPAAm-co-HMAAm)/polyurethane core-shell nanofiber mat on non-vascular nitinol stents. *Nanomedicine: Nanotechnology, Biology and Medicine, 13*(2), 527–538.
33. Perera, A. S., Zhang, S., Homer-Vanniasinkam, S., Coppens, M. O., & Edirisinghe, M., (2018). Polymer-magnetic composite fibers for remote-controlled drug release. *ACS Applied Materials and Interfaces, 10*(18), 15524–15531. https://pubs.acs.org/doi/abs/10.1021%2Facsami.8b04774.
34. Sun, X., Wang, G., Zhang, H., Hu, S., Liu, X., Tang, J., & Shen, Y., (2018). The blood clearance kinetics and pathway of polymeric micelles in cancer drug delivery. *ACS Nano, 12*(6), 6179–6192.

CHAPTER 14

Utilizing Nanotechnology for Environmental Cleaning

KULDIP DWIVEDI,[1] SHIVANI YADAV,[2] and V. K. YADAV[3]

[1]*Department of Environmental Science, Amity University Madhya Pradesh, Gwalior – 474005, Madhya Pradesh, India, E-mail: dwivedikul2012@gmail.com*

[2]*Department of Life Sciences, ITM University, Gwalior, Madhya Pradesh, India*

[3]*Division of Crop Improvement, IGFRI, Jhansi, Uttar Pradesh, India*

ABSTRACT

The rapid strides made in science and technology has increased standard of living which has contributed to the increase in waste and toxic material in the environment. Therefore, the remediation of contaminants by use of nanomaterials may be applied which will not only have less toxic effect on microorganisms but will also improve the microbial activity of the specific waste and toxic material and reduce the overall time consumption as well as reduce the overall cost. In this chapter, we have tried to briefly summarize the major types of nanomaterials, the processes exploited due to novel properties in NMs and the targeted compounds of waste and toxic materials for bioremediation. Nanotechnology has the potential to provide eco-friendly alternatives for environmental management without harming the natural environment.

14.1 INTRODUCTION

The most challenging task of the 21st century is a clean up of the environment to contaminants by ecofriendly, sustainable, economically viable, and socially acceptable technologies. Among different emerging technological options, nanotechnology has emerged as the most potential tool for combating several challenges. Nanotechnology is the design, fabrication, and utilization of materials, devices systems through control of mater in nanometer scale and exploitation of novel phenomena like physical, chemical, and biological in that scale. Nanotechnology is being used in several applications to make them environment ecofriendly. This includes cleaning up existing pollutants, improving manufacturing methods to reduce the generation of new pollution, and making cost-effective alternative energy sources. Nanotechnologies propose new opportunities for making superior materials to reduce environmental pollution. The integration of nanotechnology and material science has facilitated to develop novel nanoparticle functional materials, suitable for various applications in all the fields and particularly in the bioremediation field for the benefit of human health and well being.

The particles between the sizes of 1 and 100 nm have been considered as nanoscale particles. A eukaryote has a size of 10,000 nm, a bacteria 1,000–10,000 nm, virus 75–100 nm, protein 5–50 nm, deoxyribonucleic acid (DNA) ~2 nm (width), and an atom ~0.1 nm. NPs, the primary building blocks of many nanomaterials, are of particular interest in various studies, as the fate of NPs in aqueous environments will depend on their extraordinary properties and widespread range of applications in different scientific and industrial backgrounds. NPs have a significant impact in many scientific fields, including chemistry, electronics, medicine, biology, and material sciences [1, 2]. Synthesis of NPs using biological entities has great importance due to their unusual photoelectrochemical and electronic properties. The synthesis and assembly of nanoparticles (NPs) possess great for the development of clean, non-toxic, and environmentally acceptable 'green chemistry' procedure, the physical, material, and chemical properties of NPs are directly related to their intrinsic composition, apparent size, and unique surface structure [3, 4], so the design, the synthesis, the characterization, and the applications of nanostructures are critical aspects of the emerging field of nanomaterials. Currently, large quantities of engineered nanomaterials (ENMs) are being produced for diverse applications, and the trend is expected to increase in the future. This increases the probability that NPs will enter the environment during their production, manufacture, use, or disposal.

14.2 TYPES OF NANOPARTICLES (NPS)

NPs can be classified into different types according to their size, morphology physical and chemical properties. Important types of NPs are carbon-based NPs, ceramic NPs, metal NPs, semiconductor NPs, polymeric NPs, and lipid-based NPs.

14.2.1 CARBON-BASED NANOPARTICLES (NPS)

Carbon-based NPs include carbon nanotubes (CNTs) and fullerenes. CNTs are nothing, but graphene sheets rolled into a tube. These materials are mainly used for structural reinforcement as they are 100 times stronger than steel. CNTs can be classified into single-walled carbon nanotubes (SWCNTs) and multi-walled carbon nanotubes (MWCNTs). CNTs are unique in a way as they are thermally conductive along the length and non-conductive across the tube. Fullerenes are the allotropes of carbon having a structure of a hollow cage of sixty or more carbon atoms. The structure of C-60 is called Buckminsterfullerene, which is completely hollow from inside and forms a structure like a football. The arrangements of carbon units in CNTs are pentagonal and hexagonal, which helps in the better stability and strength of material produced utilizing such nanomaterials (NMs). CNTs have commercial applications due to high electrical conductivity, highly stable structure, high strength, and electron affinity. The available data reveals that studies on CNTs have mainly focused on animals and humans health [5, 6].

14.2.2 CERAMIC NANOPARTICLES (NPS) OR NANOCERAMICS

Ceramic NPs are inorganic solids which made up of oxides, phosphates, nitrides, carbides, and carbonates. Nanoceramic is chemically inert and possesses high heat tolerance. There are many nanoceramic materials viz. clay mineral, cement, and glass used for various applications [7]. Nanoceramics also have applications photocatalysis, photodegradation of dyes, drug delivery, and imaging. Nanomaterials possess great potential to be used in drug delivery if some characteristics of ceramic NPs like size, surface area, porosity, the surface to volume ratio, etc. are controlled and manipulated. These nanoceramics have been used significantly in biomedical application like as a drug delivery system for a number of diseases like bacterial infections, glaucoma, cancer, etc.

14.2.3 METAL NANOPARTICLES (NPS)

Precursors of different metals are utilized for the production of metals NPs. These NPs can be synthesized utilizing chemical, electrochemical, and using photochemical processes. In chemical methods, metal NPs are produced by reducing the metal-ion precursors in solution by using chemical agents, which promotes reduction. These chemical agents have the ability to absorb small molecules and have high surface energy. Metal-based NPs have wide applications in research areas, detection, and imaging of biomolecules and in environmental and bioanalytical applications. Many metal NPs viz. gold (Au), silver (Ag), and copper (Cu) are greatly utilized in different studies.

14.2.4 SEMICONDUCTOR NANOPARTICLES (NPS)

Semiconductor NPs have properties of both metals and non-metals. They are found in the periodic table in groups II-VI, III-V, or IV-VI. These type particles have wide bandgaps, which on tuning show different properties. They are used in photocatalysis, electronics devices, photo-optics, and water splitting applications. There are numerous examples of semiconductor NPs, and some of them are GaN, GaP, InP, InAs from group III-V, ZnO, ZnS, CdS, CdSe, CdTe are II-VI and silicon, and germanium are from group IV.

14.2.5 POLYMERIC NANOPARTICLES (NPS)

These types of NPs are organic in origin. Depending upon the method of production, these have different structures shaped like nanocapsular and nanospheres. A nanospheres particle has a matrix-like structure while the nanocapsular particle has core-shell morphology. In the nanocapsular type of NPs, the active compounds and the polymer are uniformly dispersed, whereas in the nanospheres type NPs the active compounds are confined and surrounded by a polymer shell. The polymeric NPs possess many advantages over the other type of NPs, which include controlled drug release, protection of drug molecules, ability to combine therapy and imaging, and specific targeting. They have applications in drug delivery, diagnostics, nutrient applications, and seed coating of agricultural and horticultural crops.

14.2.6 LIPID-BASED NANOPARTICLES (NPS)

Lipid NPs are generally round in shape with a diameter ranging from 10 to 100 nm. In this type of NPs, solid core made of lipid and a matrix containing soluble lipophilic molecules. The external core of these NPs are stabilized by surfactants and emulsifiers. These NPs have applications in the biomedical field as a drug carrier and delivery and RNA release in cancer therapy.

14.3 USE OF NANOPARTICLES (NPS) IN REMEDIATION OR ENVIRONMENTAL CLEANING

Heavy metals are the most important hazardous contaminants in the environment. The cause of huge and long term pollution. Pollution in the environment can be cleaned up (also called "remediation) using a range of techniques. The field of study that focuses on investigating the cleanup or removal of contaminants from the environment is called "environmental remediation." The remediation of contaminants by the use of existing technologies is less effective and has limitations in cleaning up of the environment. However, with the development of nanotechnology new option has been raised, which involve the use of nanomaterial for environment cleaning. Now new approaches are available both for extraction of nanomaterial from renewable resources as well as for nanoparticle synthesis. Thus, nanomaterials applied for remediation, which not only has a less toxic effect on microorganisms, but also improves the microbial activity on the specific waste and toxic material which will reduce the overall time consumption as well as reduce the overall cost.

14.3.1 UTILIZATION OF METAL-BASED NANOPARTICLES (NPS) FOR REMEDIATION

Iron NPs are considered to be the first nanoparticle to be used in an environment clean-up [8]. Iron derived polymer-coated NPs used for the removal of many heavy metals from water such as Cr and As [9, 10]. The common iron-based NMs used for environment remediation are nanosized zero-valent ion (NZVI), iron sulfide NPs, bimetallic Fe NPs, and nanosized FeO [11]. NZVI has been effectively utilized for arsenic remediation from water [12]. Iron (Fe) and nickel (Ni) based NPs have been found better alternatives for remediation of uranium. Fe and Ni-based NPs went through vacuum annealing processes and thus enhances the efficiency of uranium removal compared to iron particles of larger size [13]. The zero-valent iron particles (ZVI) technology has been successfully utilized for the decontamination of groundwater to reduce both the chlorinated hydrocarbons and metal toxicity.

Accelerated photocatalytic degradation of polychlorinated biphenyl (PCB) has been achieved by treating polluted soil titanium dioxide (TiO_2) NPs. TiO_2 MCM (Mesoporous molecular sieving) has been investigated for the photolytic degradation of the endocrine-disrupting compound, Bisphenol A (2,2-bis(4-hydroxyphenyl)propane, BPA). Titanium dioxide NPs (TiO_2-x) containing oxygen vacant sites have been synthesized using a plasma discharge method, which induces enhanced oxygen supply and accelerates the mercury (Hg) removal from the polluted source. The absorption capacity of TiO_2 NPs to adsorb contaminant metals like Pb, Cu, Cd, Ni, Zn has been used successfully [14].

Mercury is a serious threat for environment health globally due to its neurotoxin methyl mercury to biomagnify in the food chain. Silica-based NPs used in the remediation processes and it has been modified with a 2,6-pyrimidine dicarboxylic acid, and used as a sorbent successfully for the separation and preconcentration of Hg(II) ions present in trace levels. Better reusability and enhanced stability with respect to Hg ions were a positive aspect shown by the sorbent for the separation of Hg ions. The absorption equilibrium is fast, elution easy, and the absorption capacity high with the procedure being rapid and convenient [14].

Metallic NPs are also being used with the aim of removing heavy metals, such as mercury (Hg), nickel (Ni), copper (Cu), arsenic (As), cadmium (cd), chromium (Cr), lead (Pb), etc. Calcium-doped zinc oxide NPs used as a selective adsorbent for the extraction of lead ion was explored [15]. Many metal oxidized nanomaterials, along with nanosized magnetite and titanium dioxide, outcompetes the adsorption capacity of activated carbon [16]. In another study, metal hydroxide NPs have been used along with the activated carbon in order to remove arsenic and other organic pollutants [17, 18]. NPs produced from metal or metal oxides that are extensively used in wastewater treatment for eradicating the heavy metals are manganese oxides [19], copper oxides [20], cerium oxide [21], magnesium oxides [22], titanium oxides [23], silver nanoparticles

(AgNPs) [24], and ferric oxides [25]. Magnesium hydroxide nanotubes to form $Mg(OH)_2/Al_2O_3$ composites in order to remove nickel ions from contaminated water [26]. NPs of Ag, CuO, and ZnO possess good antimicrobial properties against pathogenic bacteria. The bacteria play an important role in element cycling, pollutant degradation, and plant growth. It is very important that these processes should be developed and utilized to ensure that nanomaterials are safe possible. Biofilters with long lifetime and high storage stability are very important for bioremediation processes to ensure the readiness at the time of occurrence of sudden contaminations [27]. Activated carbon is widely used for removing contaminants in public drinking system [28]. The use of ZVI materials, singly, or association with other technologies are very effective in the treatment of chlorinated solvent contamination. Iron seems to perform better if it is used singly. Due to its association with microbial dechlorination, it has been shown that the performance of iron nanomaterial is significantly enhanced. Based on the research carried, it is suggested that ZVI technologies for effective remediation warrants further application and exploration [29].

Research and application studies carried out it has been proven that NPs can potentially be utilized for the remediation of soil and groundwater. Environment cleaning technologies can classify as adsorptive or reactive and as in-situ or ex-situ [30]. Available nanomaterial-based approached deployed for remediation can be categorized in different groups based on the process utilized in the remediation (Table 14.1).

14.4 CONCLUSION

Nanotechnology has the potential to revolutionize existing technologies used in various sectors, including pollution control. Nanotechnology emerging as plays a major role in the development of new products to substitute existing production processes, with improved performance, resulting in potential environmental and cost savings. Reduced consumption of materials is also beneficial. Moreover, nanotechnology provides the potential to organize and develop production processes in a more sustainable way, eventually as close to a zero-emissions approach as possible. Nanotechnology may provide effective solutions for many pollution-related problems such as heavy metal contamination, adverse effects of chemical pollutants, oil pollution, and so on. Nanotechnology

TABLE 14.1 Use of Nanoparticles in Remediation

Process exploited	Target Compounds	Nanomaterials Used	Some of Novel Properties	References
Adsorption	Heavy metals, organic compounds, arsenic, phosphate, Cr, Hg, PAHs, DDT, Dioxin	Iron oxides, carbon nanotubes (CNTs)	High specific surface area and assessable adsorption sites, selective, and more adsorption sites	[31]
Photocatalysis	Organic pollutants, NOX, VOCs, Azo dye, Congo red dye, 4-chlorophenol, and Orange II, PAHs	TiO_2, ZnO, Species of iron oxides (Fe III, Fe_2O_3, Fe_3O_4)	Photocatalytic activity in the solar spectrum, low human toxicity, high stability and selectivity	[32]
Redox reactions	Halogenated organic compounds, metals, nitrate, arsenate, oil, PAH, PCB	Nanoscale zero-valent iron, calcium peroxide	Electron transfers such as photosynthesis, respiration, metabolism, molecular signaling	[33]
Disinfection	Diamines, phenols, formaldehyde, hydrogen peroxide, silver ions,	Nanosilver/titanium dioxide (Ag/TiO_2) and CNTs	Strong antimicrobial activity, low toxicity, and cost, high chemical stability	[34]
Membranes	Chlorinated compounds, polyethylene, 1,2-dichlorobenzene, organic, and inorganic solutes	NanoAg/TiO_2/Zeolites/Magnetite and CNTs	Strong antimicrobial activity, hydrophilicity, low toxicity to humans	[35]

could provide eco-friendly alternatives for environmental management without harming the natural environment. However, while exploiting the potential benefits of nanomaterials in water treatment processes and combating other forms of pollution, it also suggested giving adequate attentions towards concerns that have been raised regarding their potential human and environmental toxicity.

KEYWORDS

- **carbon nanotubes**
- **deoxyribonucleic acid**
- **multi-walled carbon nanotubes**
- **nanoparticles**
- **nanosized zero-valent ion**
- **single-walled carbon nanotubes**

REFERENCES

1. Schmid, G., Bäumle, M., Geerkens, M., Heim, I., Osemann, C., & Sawitowski, T., (1999). Current and future application of nanoclusters. *Chem. Soc. Rev., 28*, 179–185.
2. Boisselier, E., & Astruc, D., (2009). Gold nanoparticles in nanomedicine, preparations imaging diagnostics, therapies and toxicity. *Chem. Soc. Rev., 38,* 1759–1782.
3. Mirkin, C. A., Letsinger, R. L., Mucic, R. C., & Storhoff, J. J., (1996). A DNA based method for rationally assembling nanoparticles into macroscopic materials. *Nature, 382*(6592), 607.
4. Zhong, W., (2009). Nanomaterials in fluorescence-based biosensing. *Anal. Bioanal. Chem., 394,* 47–59.
5. Ke, P. C., Lin, S., Reppert, J., Rao, A. M., & Luo, H., (2011). Uptake of carbon-based nanoparticles by mammalian cells and plants. In: Sattler, K. D., (ed.), *Handbook of Nanophysics: Nanomedicine and Nanorobotics* (pp. 1–30). CRC Press, New York.
6. Tiwari, D. K., Dasgupta-Schubert, N., Villaseñor-Cendejas, L. M., Villegas, J., Carreto-Montoya, L., & Borjas-García, S. E., (2014). Interfacing carbon nanotubes (CNT) with plants: Enhancement of growth, water and ionic nutrient uptake in maize (Zea Mays) and implications for nanoagriculture. *Appl. Nanosci., 4,* 577–591.
7. Balasubramanian, S. G., & Balasubramanian, A., (2017). biomedical applications of ceramic nanomaterials. *Rev., 8*(12) 4950–4959.
8. Tratnyek, P. G., & Johnson, R. L., (2006). Nanotechnologies for environmental cleanup. *Nano Today, 1*, 44–48.
9. Abdollahi, M., Zeinali, S., Nasirimoghaddam, S., & Sabbaghi, S., (2015). Effective removal of As(III) from drinking water samples by chitosan-coated magnetic nanoparticles. *Desalination and Water Treatment, 56*, 2092–2104.

10. Mu, Y., Ai, Z., Zhang, L., & Song, F., (2015). Insight into core–shell dependent anoxic Cr(VI) removal with Fe @ Fe2O3 nanowires: Indispensable role of surface bound Fe(II). *ACS Applied Materials and Interfaces, 7*, 1997–2005.
11. Ludwig, R. D., Su, C., Lee, T. R., Wilkin, R. T., Acree, S. D. Ross, R. R., & Keeley, A., (2007). In situ chemical reduction of Cr(VI) in groundwater using a combination of ferrous sulfate and sodium dithionite: A field investigation. *Environmental Science and Technology, 41*, 5299–5305.
12. Kanel, S. R., Greneche, J. M., & Choi, H., (2006). Arsenic (V) removal from groundwater using nanoscale zerovalent iron as a colloidal reactive barrier material. *Environmental Science and Technology, 40*(6), 2045–2050.
13. Faulkner, D. W. S., Hopkinson, L., & Cundy, A. B., (2005). Electrokinetic generation of reactive iron-rich barriers in wet sediments: Implications for contaminated land management. *Mineralogical Magazine, 69*, 749–757.
14. Zhang, L, Chang, X., Hu, Z., Zhang, L., Shi, J., & Gao, R., (2010). *Microchim Acta Selected Solid Phase Extraction and Preconcentration of Mercury (II) from Environmental and Biological Samples Using Nanometer Silica Functionalized by 2,6-Pyridine, 168*, 79–85.
15. Khan, S. B., Marwani, H. M., Asiri, A. M., & Bakhsh, E. M., (2015). Exploration of calcium doped zinc oxide nanoparticles as selective adsorbent for extraction of lead ion. *Desalination and Water Treatment*, doi: 10.1080/19443994.2015.1109560.
16. Mayo, J. T., Yavuz, C., Yean, S., Cong, L., Shipley, H., Yu, W., Falkner, J., Kan, A., Tomson, M., & Colvin, V. L., (2007). The effect of nanocrystalline magnetite size on arsenic removal. *Science and Technology of Advanced Materials, 8*(1&2), 71–75.
17. Hristovski, K. D., Nguyen, H., & Westerhoff, P. K., (2009). Removal of arsenate and 17-ethinyl estradiol (EE2) by iron (hydr) oxide modified activated carbon fibers. *Journal of Environmental Science and Health Part A—Toxic/Hazardous Substances and Environmental Engineering, 44*(4), 354–361.
18. Hristovski, K. D., Westerhoff, P. K., Moller, T., & Sylvester, P., (2009). Effect of synthesis conditions on nano-iron (hydr) oxide impregnated granulated activated carbon. *Chemical Engineering Journal, 146*(2), 237–243.
19. Gupta, K., Bhattacharya, S., Chattopadhyay, D., Mukhopadhyay, A., Biswas, H., Dutta, J., Ray, N. R., & Ghosh, U. C., (2011). Ceria associated manganese oxide nanoparticles: Synthesis, characterization and arsenic (V) sorption behavior. *Chemical Engineering Journal, 172*, 219–229.
20. Goswami, A., Raul, P. K., & Purkait, M. K., (2012). Arsenic adsorption using copper (II) oxide nanoparticles. *Chemical Engineering Research and Design, 90*(9), 1387–1396.
21. Cao, C. Y., Cui, Z. M., Chen, C. Q., Song, W. G., & Cai, W., (2010). Ceria hollow nanospheres produced by a template-free microwave-assisted hydrothermal method for heavy metal ion removal and catalysis. *Journal of Physical Chemistry, 114*, 9865–9870.
22. Gao, C., Zhang, W., Li, H., Lang, L., & Xu, Z., (2010). Controllable fabrication of mesoporous MgO with various morphologies and their absorption performance for toxic pollutants in water. *Crystal Growth and Design. 8,* 3785–3790.

23. Luo, T., Cui, J., Hu, S., Huang, Y., & Jing, C., (2010). Arsenic removal and recovery from copper smelting wastewater using TiO2. *Environmental Science and Technology, 44*, 9094–9098, 2010.
24. Fabrega, J., Luoma, S. N., Tyler, C. R., Galloway, T. S., & Lead, J. R., (2011). Silver nanoparticles: Behavior and effects in the aquatic environment. *Environmental International, 37*, 517–531.
25. Feng, L., Cao, M., Ma, X., Zhu, Y., & Hu, C., (2012). Super paramagnetic high-surface area Fe_3O_4 nanoparticles as adsorbents for arsenic removal. *Journal of Hazardous Materials, 217&218*, 439–446.
26. Zhang, S., Cheng, F., Tao, Z., Gao, F., & Chen, J., (2006). Removal of nickel ions from wastewater by Mg(OH)2/MgO nanostructures embedded in Al2O3 membranes. *Journal of Alloys and Compounds, 426*, 281–285.
27. Gajjar, P., Pettee, B., Britt, D. W., Huang, W., Johnson, W. P., & Anderson, A. J., (2009). Antimicrobial activities of commercial nanoparticles against an environmental soil microbe, pseudomonas putida KT2440. *Journal of Bio. Eng., 3*, 9–22.
28. Brar, S. K., Verma, M., Tyagi, R. D., & Surampalli, R. Y., (2010). Engineered nanoparticles in wastewater and wastewater sludge. *Waste Manag., 30*, 504–520.
29. Duran, N., (2008). *Use of Nanoparticles in Soil-Water Bioremediation Processes, 5th International Symposium ISMOM* (pp. 1–6). Chile.
30. Comba, S., Molfetta, A. D., & Sethi, R., (2011). Aqueous contaminant removal by metallic iron: Is the paradigm shifting. *Water Air Soil Poll, 215*, 595–607.
31. Bhaumik, M., Maity, A., Srinivasuc, V. V., & Onyango, M. S., (2012). Removal of hexavalent chromium from aqueous solution using polypyrrole-polyaniline nanofibres. *Chem. Eng. J., 181&182*, 323–333.
32. Khedr, M., Abdelhalim, K., & Soliman, N., (2009). Synthesis and photocatalytic activity of nano –sized iron oxide. *Mater. Lett., 63*, 598–601.
33. Nowack, B., (2008). In: Krug, H., (ed.), *Nanotechnology* (pp. 1–15). Wiley-VCS Verlag GmbH & Co, Weinheim.
34. Amin, M. T., Alazba, A. A., & Manzoor, U., (2014). Removal of pollutants from water/waste water using different types of nanomaterials. *Adva. Mater. Sci. Eng.* article ID 825910.
35. Donnell, G. M., & Russell, A. D., (1999). A nanobioremediation technologies for environmental cleanup. *Clin. Microbiol. Rev., 12*, 147–179.

CHAPTER 15

Implementation of Nanotechnology in Agriculture System: A Current Perspective

RAGHVENDRA SAXENA,* MANISH KUMAR, and RAJESH SINGH TOMAR

Amity Institute of Biotechnology, Amity University Madhya Pradesh, Gwalior, India

**Corresponding author. E-mail: rsaxena@gwa.amity.edu*

ABSTRACT

Nanotechnology is the emerging technology in the present scenario and offers wide spectrum opportunities to the progressive world. Agriculture is one of the areas which fulfill the food demand of a growing global population, but this sector is facing several challenges that can be addressed and resolved. In the last decade, nanobiotechnology has emerged as a promising technology to address the issues of agricultural sustainability. The present chapter is the compilation of several opportunities offered by nanobiotechnology to boost the agriculture sector by addressing the current issues of agricultural sustainability. Applications of engineered nanoparticles (NPs), metal-based, or metal oxide-based NPs have shown the potential to improve plant growth and crop production. The chapter is further elaborated on the agronomic applications of nanobiotechnology. It offers a huge opportunity to engineer fertilizer development with desires chemical composition that improves nutrient use efficiency and enhances crop productivity. Agronomic applications of nanobiotechnology allow better resource utilization by controlled release of agrochemicals for target-specific delivery. Applications of genetic engineering for crop improvement using NPs for efficient genetic delivery systems further change gene transformation scenarios in plants, which are limited by several restrictions. The improved understanding of nanobiotechnology with reference to plants and their interaction with NPs helps, development of nanoparticle, nanobiosensors, which could be the major steps toward agriculture sustainability.

15.1 INTRODUCTION

Nanotechnology is an interdisciplinary research field that utilizes multiple streams of sciences, including physics, chemistry, medicine, material science, pharmaceutical, agriculture, etc. This technology manipulates the changing properties of the matter at nanoscale to design the tools and applications. Recently nanotechnology has been considered as a sixth important emerging technology in the field of science that could provide many fold boosts in the agriculture sector to cope with growing food demand and agriculture sustainability. However, developments of nanotechnology in the agriculture sector need to get its wide applications and exists in the juvenile phase [1, 2]. In the last decade, nanotechnology has created a wide interest in the scientific community to put serious efforts to develop nanotechnology-based applications in the agriculture sector. The term "nanobiotechnology" was the first time used by Lynn W. Jelinski, who was a biophysicist at Cornell University, USA. The latest updates in fast-growing nanotechnology lead to the development of several innovative and potential applications in the agriculture sector. Nanobiotechnology combines nanotechnology with the biological system to develop innovative nanofabricated applications for biological research and precision agriculture [3]. Nanobiotechnology utilizing the unique properties of materials exhibited at nanoscale size makes it the most suitable candidate to develop potential applications in order to boost the agriculture sector, which includes agricultural production, crop protection, food processing, food safety through improved packaging and nanosensor, etc. [4–6]. The term nanobiotechnology is derived from the combination of words nano and biotechnology. Nanotechnology offers technological options for the investigation and transformation of biological systems, whereas biology provides biological components to nanotechnology. Therefore nanobiotechnology opens up immense opportunity to develop innovative nanofabricated devices and tools based on bio structured machines [7–9].

Nanotechnology explores the wide-area and opens large opportunities for diverse applications in the biotechnology and agricultural sector to bolster crop productivity and enhance efficacy with minimum inputs of resources and cost. Moreover, the potential benefits of nanotechnology need to be explored by the development of nanoparticles (NPs), nano-fertilizers, nano pesticides, nano herbicides, etc. The development of nanobiosensors for controlled delivery systems and effective release agrochemicals could be exploited further in the agricultural sector for precision [10–12]. Nanotechnology offers multidisciplinary applications that control unique properties of matter at nanoscale and exploiting these properties nanotechnology may benefit the agriculture system in several ways including, development of nano-fertilizers encapsulated in NPs which lead to sustained and controlled release of fertilizers in small quantities, development of nanoparticle-based genetic material delivery system for genetic transformation and crop improvement purposes, development of nanosensors, nanopesticides, nanoherbicides for effective weed control and disease management to enhance crop productivity, agricultural sustainability and food security [13]. Nanobiotechnology is important in addressing issues of agricultural sustainability, climate change, adverse environmental impact, health hazards, food security, etc. through several potential applications, especially in agriculture, which can manage plant growth, effective disease management and stress management for crops to further augment crop production [14–16]. NPs effect on plants through several changes occur at the morphological, physiological, and biochemical level; therefore it is necessary to understand the effective role of nanoparticle in plants and their subsequent use in the development of a tool to ameliorate agriculture sector [17]. In the last decade, nanobiotechnology witnessed huge progress and build-up of more understanding of diverse applications in the agriculture sector. Nanobiotechnology will hold key positions and an important economic driving force that will not only benefit the consumer but also help farmers with no adverse effects on humans and the environment.

15.2 NANOPARTICLES (NPS)

Nanoparticles or nano-scale particles (NPs/NSPs) are small molecular aggregates having dimensions between 1 and 100 nm [18]. These NPs acquire entirely unique and diverse physicochemical properties such as size, enhanced reactivity, broader surface area, flexible pore size, and morphology as compared to their bulk material [19]. The engineered NPs are developed and designed to acquire the designated properties, which are otherwise not present in their bulk compound. NPs impart high surface energy and high surface to volume ratio, which increases their reactivity and other biochemical attributes, such unique features of NPs may cause different behavior and impact than their bulk counterparts [20, 21]. NPs are ubiquitous in nature, and plants have evolved with exposure to them [22]. NPs are not strangers to the environment; they exist naturally in several forms of clays, minerals, and even in the form of certain bacterial products. Even certain plants also exhibited the property of producing natural mineralized NPs. Studies revealed that iron oxide nanoparticles (IONPs) occupy an important position as a part of naturally occurring- ring NPs among other NPs exits naturally [23]. There are studies that suggested that plant and soil microbes exhibit a natural tendency to produce IONPs [22, 24, 25]. Plants also attributed inherent capabilities to produce naturally mineralized nanomaterials (NMs) under certain conditions that are necessary to their growth and development, as revealed under some observations [26].

Recently Palmqvist et al. [27] synthesized and characterized Yttrium doping-stabilized γ-Fe_2O_3 NPs, these maghemite NPs acts as nanozymes, which could be a potential candidate as plant fertilizer and it can be used against drought stress in plants to mitigate its impacts. The study revealed the significant increase in drought tolerance in plants under the influence of maghemite NPs by modulating oxidative defense enzymes system in *Brassica napus*; moreover, it also helped in reducing oxidative stress to a significant level [27]. Since metals and metal oxides NPs acquire a high surface to volume ratio, they acquire additional properties, which include enhanced activity and increased bio-availability and other biochemical activities [20]. So it is very imperative the nanobiotechnology offers potential applications in current agricultural practices to address several agricultural issues, i.e., crop productivity, host defense again pests, effective nutrient management, development of tolerance against biotic as well as abiotic stress, etc. The engineered NPs are designed from the combination of different materials to achieve designated properties; the current research scenario is now more focused on the synthesis of novel engineered NPs using several strategies. Development of NPs from metal and metal oxides has proved their importance as these NPs impart unique physicochemical properties and could be utilized in the development of several applications important in the agriculture system to exploiting plant-nanoparticle interaction [28, 29].

In another observation, which revealed that NPs might vary or change their properties based on their composition and environment, therefore the same NPs response differently to the plants, for example, capped nanoparticle impart its effect on plants differently as compared to uncapped NPs [30]. Agriculture is one of the areas where plenty of opportunities exist for nanobiotechnology; therefore, several engineered NPs could be developed and utilized in agriculture system as nano-fertilizers, nanopesticides or as nanocarriers for pesticides and fertilizers, etc. [31, 32]. NPs can be synthesized from the wide range of heterogeneous metals for industrial applications, however, at present silver, zinc, copper, aluminum, iron, nickel, etc. and their metal oxides NPs are most commonly used NPs by the industries, therefore they are most preferred to study for their effect on plants [33].

In recent years, chitosan emerged as one of the most potential nanostructured polymers in the agriculture sector as an effective carrier for agrochemicals and micronutrients. Chitosan is a biodegradable polymer that is derived from chitin by deacetylation [34]. The low concentration and effective release of agrochemicals are one of the prime concerns of agriculture sustainability as these agrochemicals are highly stable and persistent in the soil, which may lead to toxicity [35]. Therefore, to enhance efficacy and effective release of herbicides (Imazapic and Imazapyr), encapsulated in chitosan NPs was observed to be an effective tool in this direction [36].

15.3 BIOGENIC OR GREEN SYNTHESIS OF NANOPARTICLES (NPS)

With the growing information of nanotechnology, the synthesis of nanomaterial of commercial relevance from different metals and metal oxides is vital areas of research. In this direction, two basic approaches are commonly adopted tor synthesis nonmaterial or NPs. The one is referred to as a top-down approach, which depends on size reduction from bulk materials, whereas another one is referred to as the bottom-up system approach, which involves material synthesis from atomic (according to

Royal Society and Royal Academy of Engineering) [1]. NPs synthesis, either directly in the presence of chemicals or in the presence of biological components holds the key importance without compromising the efficacy. The synthesis of several metal NPs, including Au, Ag, Fe, Pt, Ti, Zn, Mg, etc. have been successfully synthesized so far by using different approaches [37]. A wide range of metal oxide NPs, i.e., ZnO, TiO_2, Al_2O_3, FeO, Fe_2O_3, etc., fullerenes, carbon nanotubes (CNTs), quantum dots (QDs), etc. have further broadened the scope of nanotechnology to develop the wide range of applications [38]. Metal and metal oxide NPs are synthesized by the addition of reducing/oxidizing or precipitant agents, respectively [39]. It is important to understand the method through which NPs are being synthesized because the properties of NPs governed by several factors, which determine the reactivity of nanoparticle, include size, surface properties, composition, stability, etc. [32]. In the last decade, the emphasis was given on the development of green methods for the synthesis of NPs, which are not only cost-effective but also eco-friendly options. Among the green methods, microbes and plants are the most convenient and suitable candidates for large-scale biosynthesis of NPs. NPs synthesis using plant extracts exhibit better stability and can be synthesized rapidly as compared to physical and chemical methods. Therefore, a wide range of plant extracts needs to be exploited. Moreover, NPs synthesized using plant extract not only cost-effective but also benign to the environment, most importantly, the biological approach imparts more stability as compared to those synthesized using physical and chemical methods [40]. The biosynthesis of wide range NPs using plants or microbes as a biological component has switched the entire scenario of nanoparticle synthesis in a positive direction to adopt an eco-friendly approach (Green chemistry), where no strong, toxic, and costly chemicals are used [41].

Commercial and large scale production of NPs using plant extract involved bioreduction of NPs by the combination of several biomolecules in the present plant extract including enzymes, vitamins, phenolics, proteins amino acid, polysaccharides, and certain organic acids, etc. which play a vital role in nanoparticle production [42]. There are several applications of NPs in medicine and the pharmaceutical sector. Gold and silver nanoparticles (AgNPs) are the most important and widely used in nanobiotechnology in these sectors. So several studies reported the synthesis and characterization of silver and gold nanoparticles (AuNPs) using different plant extracts as reducing agent i.e., *Pelargonium graveolens*,

Azadirachta indica, Cymbopogon flexuosus (lemongrass), *Coriandrum sativum, Aloe vera, Tamarindus indica* (tamarind), these NPs were well characterized for their size and these NPs assemble and acquired variety of shapes (morphological shapes), i.e., triangular (nanotriangles), spherical, decahedral, and were highly stable at room temperature [43–46]. So the recent advances in the green synthesis of NPs using plant extract have open immense opportunity to explore the wide range of plants grown in diverse climatic conditions for eco-friendly synthesis of NPs.

15.4 NANOBIOTECHNOLOGY IN NUTRITIONAL SUSTAINABILITY TO CROP PLANTS

The most vital issue in recent years regarding food production is to increase food production to address the pressing food demand globally under limited resources and changing climate scenarios. Increasing the efficiency and minimum use of agrochemicals like fertilizers, herbicides, and pesticides, and other agrochemicals without adversely affecting the environment are in great interest of agriculture sustainability.

Effective nutritional uptake by the crop is one of the vital factors that control plant growth and productivity. However, the excessive and indiscriminate application of chemical fertilizers in the agriculture field increased productivity but also raised the environmental issues, which included decreased soil fertility, accumulation of fertilizers in the soil leading to perturb soil harmony, etc. which adversely affects the crop productivity. However, in the recent years, nano-fertilizers extends new opportunity in the agriculture sector to enhance crop productivity by striking a nutritional balance to plants, which are an essential component for proper growth and better productivity without compromising environmental issues. Several nanomaterials were tested for their impact on seed germination, seedling growth, plant growth, pathogen attack, etc., which can improve crop productivity. The studies on the uptake of several metals or metal oxide NPs, i.e., ZnO, TiO_2, FeO Al_2O_3, etc. by the plants revealed a significant positive effect on plant growth. Apart from NPs, CNTs, QDs, nanorods also exhibit a unique set of properties that offers potential applications in the development of nanodevices and nanostructures that can be used in agriculture to enhance crop productivity and crop protection. For instance, carbons nanotubes (CNTs) can be used in targeted delivery of agrochemicals, thus offer efficient use of resources and prevent excess chemicals released into the environment [47]. QDs exhibit unique spectral features that enables it excellent fluorescent material that can be used to develop nanodevices for live bioimaging and biosensing in plants to monitor the physiological processes [48].

15.5 NANOBIOTECHNOLOGY IN CROP PROTECTION AND DISEASE MANAGEMENT IN PLANTS

Nanotechnology offers great potential in the agricultural sector for precise and effective agriculture management practices by augmenting the efficacy of the management system and conservation of resource inputs in crops [49]. Disease and pathogen attacks are major limiting factors for crop development and productivity. Agrochemicals like pesticides, insecticides are most widely applied in agriculture to improve crop yield. Conventional methods and approaches for the application of pesticides in the agricultural felids suffer several drawbacks such as uneven distribution and application of more than required quantity and accumulation in the soil. However, nanotechnology helps in preventing the excessive use of chemical pesticides by offering the controlled delivery system for pesticides/insecticides/fertilizers for effective agricultural management. Nano-encapsulation is the most potent tool that could offer the best control over the effective release of the agrochemical, thus provides better management in plant host defense against pests and insects. The encapsulation of important agrochemicals like pesticides, fungicides, or nematicides with suitable NPs provides an effective solution for sustained release of agrochemical offers an effective tool towards pest's management and nutrient management as well. Moreover, controlled the release of agrochemicals into the soil also effectively absorbed by plants and do not accumulate in the soil to the toxic level, therefore make a more environmental-friendly approach. Nanotechnology is also helpful in releasing a very small quantity of agrochemicals including pesticides, herbicides, fungicides, etc. at the target site leading to effective uptake; therefore it offers better control and effective management of agrochemicals and reduces unused access accumulation of agrochemicals thus making it more environmental-friendly and cost-effective [49]. The focus on the development of nanoformulated agrochemicals intended to enhance the solubility of less soluble agrochemical to increase their wide impact and protection against biotic stress [50]. In the recent year's development of nanoencapsulated formulation for effective and sustained release of pesticides, fungicides, and herbicides hold the key concern in crop protection management practices, which further broaden the crop protection scenario in the agriculture sector and minimizing crop loss and the adverse impact on the environment [51].

15.6 NANOBIOTECHNOLOGY IN ABIOTIC STRESS TOLERANCE TO PLANTS

Abiotic stress, which includes salinity, drought, heat, heavy metal toxicity, flood, etc. are the major limiting factors that affect plant growth and productivity to a significant extent leading to declining in crop yield. Realizing the frequent incidences of diverse stresses and changing climate scenarios globally in recent years, sincere efforts were made to apply nanotechnology in the agriculture sector to cope with the deleterious effect abiotic as well as biotic stress on crop yield to maintain food availability. Nanobiotechnology could be applied in the development of tolerant crops or by mitigating the adverse effect on stress on plants [21]. Several studies in the last decade suggested the potential role of NPs in crop protection against abiotic stress. The application of different concentrations of silicon nanoparticles (SNPs at 0, 10, 50, and 100 mg L^{-1}) on hawthorns (*Crataegus sp.*) seedlings revealed that SNP played in a positive role in maintaining the important physiological and biochemical attributes under drought stress [52]. In the last decade, there are several metals or metal oxide-based NPs, i.e., AgNPs in wheat under salinity stress [53]. CuNPs, FeNPs, ZnNPs in wheat infected with eyespot causal agents [54]. ZnNPs and CuNPs (Zinc and Copper) NPs on wheat under drought stress [55], TiO2NPs on spinach seedling growth [56], ZnONPs on lupine (*Lupinus termis*) plants under salinity stress, AgNPs nonpriming in rice seed revealed the enhanced seed germination and enhanced antioxidant system [57], were synthesized and tested for their efficacy in seed germination, seedling growth and under stress tolerance in various crops wheat, spinach, lupin, Rice etc when seed were pre-treated with these NPs. Although the mechanism of NP induces seed germination is not well understood yet. Some studies suggested that TiO_2 and SiO_2 NPs have shown the potential to improve seed germination and enhanced the antioxidant system in *Glycine max* and onion seeds [58, 59].

15.7 NANOBIOTECHNOLOGY IN GENETIC ENGINEERING AND CROP IMPROVEMENT

Genetic engineering towards crop improvement in plants is an important tool to address agricultural sustainability and climate change etc. The development of improved verities of the desired trait through genetic engineering is important to meet inflating food demand. Therefore, it is a prerequisite to have an effective genetic transformation tool and gene delivery system for plants. Although there are several methods of gene delivery system available for plants especially biolistic method and *Agrobacterium*-mediated gene delivery system, electroporation, etc. But these methods have their own limitations which restrict their wide spectrum application, for example, less efficiency, tissue damage, effective only in certain types or species of plants, etc. Gene delivery in the plants is more complex as compared to the animal system due to the presence of a multilayered cell wall, which additionally puts a hindrance in the gene delivery system [60]. In recent years' development of nanobiotechnology also opens an opportunity for potential technology development in the gene delivery system with strong efficacy in the plant system. The role of NPs as an instrument for the gene delivery system in plants is gaining much attention in recent years. There are reports available on the role of NPs as drug and gene delivery vehicles in an animal system [61]. So far, several applications of NPs are studied in the plants for agrochemical and fertilizer delivery systems [62].

Recent studies indicated that most of the NPs adsorb on the surface of biomolecules and get access into the cell. Therefore, studies on DNA conjugated NPs have taken much concern with the development of suitable strategies for nanoparticle surface modification to improve the efficacy and stability of nanoparticle as the delivery system to transform plants [63]. Recently calcium phosphate nanoparticle (CaPN) based gene delivery system has shown better genetic transformation frequency in *Cichorium intybus* L. plants and offers a better choice over conventional methods of gene transformation to develop crop plants with desired agronomic traits [64]. Mou, Chang, et al. [65] studied better gene transformation in

Arabidopsis thaliana roots using the functionalized mesoporous silica nanoparticles (MSNs). The results indicated that MSNs act as an effective gene delivery vehicle without the need for any mechanical force [65].

15.8 NANOBIOSENSORS IN PRECISION AGRICULTURE

Nanotechnology has revolutionaries several sectors of industries, such as the food and agriculture sectors. Food and agriculture sectors are now more sensitive to technology development. These sectors which derive the economic growth are now well-adopting technology in developing crops, integrated pest management, precision agricultural practices, food processing, food packaging, livestock development, etc. In the last decade, nanosensors have emerged as a promising tool for the applications in agriculture and food production. Development of nanosensors and their corresponding biological version as nanobiosensors have opened the new heights in the food and agriculture sector by providing highly sensitive analytical tools alternative to conventional chemical and biological sensors. The nanobiosensors offers have several advantages over conventional sensors. Nanobiosensors exhibit high sensitivity due to strong signal amplification and selectivity. Nanosensors have revolutionized the agriculture sector by providing real-time sensing assets for effective and timely management practices in precision farming. Nanosensors are advantageous because of their low cost and portability. Nanobiosensors offer several detection measures like food contamination, pathogens, heavy metals, pollutants [66], monitoring of physicochemical and biological attributes of soil like temperature, humidity, pH, etc. [67].

Recently, several micro nano-based systems developed as a project of the European Commission (2015) and applied for smart agri-food systems. Currently, nanobiosensors offers smart technology in the precision pharming like crop monitoring, sustained, and effective release of nutrients and other agrochemicals in small quantity, nutrient immobilization, monitoring soil pH, soil moisture environmental changes and monitoring of interaction between plant and pathogens, etc. This tiny network of wireless nanosensors can timely sense and provide early warning according to changing conditions leading to efficient agricultural management. The network of nanobiosensors in the field helps the effective and efficient utilization of resources, whether it is water, nutrients, fertilizers, pesticides, and herbicides insecticides. Most importantly, these nanobiosensors raise the automatic alert in real-time conditions, therefore help farmers in taking appropriate, timely measures which help not only better resource utilization and enhanced crop protection and production but also reduce agricultural cost, agrochemical wastage and their environment accumulation [68]. In the recent development, nanobarcodes and nanoprocessing potentially offer the best and early detection of biotic as well as abiotic stresses, which include pathogen attack, insect detection, the onset of disease, chemical, and toxic contaminants, etc., to adopt quick response management to prevent crop loss [69]. Aptasensors are biosensors that are made up of a combination of biocomponent (single-stranded DNA or peptides molecules) and nanomaterial, which can be of different types of metal NPs, carbon NPs, magnetic NPs, etc. These aptasensors are very sensitive and target specific, which can detect their target (microbes, viruses proteins precisely and send early warning signals in order to take appropriate preventive measures. Aptasensors can also detect pesticides, insecticides, antibiotics drugs, and their residues, which are the main concern in today's scenario in the food industry [22, 70, 71]. Technological innovations and involvement of nanotechnology in agriculture sectors help farmers to be great extant by enabling them with better agriculture management practices and effective control over agricultural resources and different phases of crop cultivation [69].

15.9 CONCLUSION

The agriculture sector is the principal supplier of food for consumption; therefore, more focus on nanotechnology research in the agricultural sector is become to be necessary even a key factor for the sustainable developments in the present scenario. The involvement of nanotechnology in the agriculture sector will extend great help to the farmers and enabling them with better agriculture management practices. The development of nano-based fertilizers, nanopesticides, nanoherbicides, nanosensors, and smart delivery systems for agrochemicals offers better resource utilization with minimum loss. The development of green synthesis approaches for nanomaterials/NPs, and their subsequent use of the agriculture sector further strengthens the eco-friendly approach; however, its long term environmental impact needs to be evaluated thoroughly. In the present scenario, development of advanced nanotechnologies tools and techniques can improve the way. Agriculture is seen, and in the long term, it may provide an economic development route for sustainable agriculture globally.

KEYWORDS

- **abiotic stress**
- **mesoporous silica nanoparticles**
- **nanobiosensors**
- **nano-fertilizers**
- **nanoparticles**
- **nanotechnology**
- **sustainable agriculture**

REFERENCES

1. Abobatta, W. F., (2018). Nanotechnology applications in agriculture. *Acta Scientific Agriculture, 2*(6), 99–102.
2. Mousavi, S. R., & Rezaei, M., (2011). Nanotechnology in agriculture and food production. *Journal of Applied Environmental and Biological Sciences, 1*(10), 414–419.
3. Robinson, D. K. R., Morrison, M., et al., (2009). *Nanotechnology Developments for the Agri-Food Sector Report.*
4. Ghormade, V., Deshpande, M. V., & Paknikar, K. M., (2011). Perspectives for nanobiotechnology enabled protection and nutrition of plants. *Biotechnol. Adv., 29*, 792–803. doi: 10.1016/j.biotechadv.2011.06.007.
5. Perez-de-Luque, A., Diego, R., et al., (2009). Nanotechnology for parasitic plant control. *Pest Manag. Sci., 65*, 540–545.
6. Torney, F., Trewyn, B. G., Lin, V. S. Y., Wang, K., et al., (2007). Mesoporous silica nanoparticles deliver DNA and chemicals into plants. *Nat. Nanotechnol., 2*, 295–300.
7. Fortina, P., Kricka, L. J., Surrey, S., Grodzinski, P., et al., (2005). Nanobiotechnology: The promise and reality of new approaches to molecular recognition. *Trends Biotechnol., 23*, 168.
8. Lowe, C. R., et al., (2000). Nanobiotechnology: The fabrication and applications of chemical and biological nanostructures. *Curr. Opin. Struct. Biol., 10*, 428.
9. Bohr, M. T., et al., (2002). Nanotechnology goals and challenges for electronic applications. *IEEE Trans. Nanotechnol., 1*, 56.

10. Oliveira, J. L., Campos, E. V., Bakshi, M., Abhilash, P. C., & Fraceto, L. F., (2014). Application of nanotechnology for the encapsulation of botanical insecticides for sustainable agriculture: Prospects and promises. *Biotechnol. Adv., 32*, 1550–1561. doi: 10.1016/j.biotechadv.2014.10.010.
11. Campos, E. V., Oliveira, J. L., & Fraceto, L. F., (2014). Applications of controlled release systems for fungicides, herbicides, acaricides, nutrients, and plant growth hormones: A review. *Adv. Sci. Eng. Med., 6*, 373–387. doi: 10.1166/asem. 2014.1538.
12. Grillo, R., Abhilash, P. C., & Fraceto, L. F., (2016). Nanotechnology applied to bio-encapsulation of pesticides. *J. Nanosci. Nanotechnol., 16*, 1231–1234. doi: 10.1166/jnn.2016.12332.
13. Prasad, R., Bhattacharyya, A., & Nguyen, Q. D., (2017). Nanotechnology in sustainable agriculture: Recent developments, challenges, and perspectives. *Front. Microbiol., 8*, 1014. doi: 10.3389/fmicb.2017.01014.
14. Maruyama, C. R., Guilger, M., Pascoli, M., Bileshy-José, N., Abhilash, P. C., Fraceto, L. F., et al., (2016). Nanoparticles based on chitosan as carriers for the combined herbicides imazapic and imazapyr. *Sci. Rep., 6*, 19768. doi: 10.1038/srep 19768.
15. Mishra, S., Keswani, C., Singh, A., Singh, B. R., Singh, S. P., & Singh, H. B., (2016). Microbial nanoformulation: Exploring potential for coherent nano-farming. In: Gupta, V. K., Sharma, G. D., Tuohy, M. G., & Gaur, R., (eds.), *The Handbook of Microbial Bioresourses* (pp. 107–120). London: CABI.
16. Mishra, S., Keswani, C., Abhilash, P. C., Fraceto, L. F., & Singh, H. B., (2017). Integrated approach of agri-nanotechnology: Challenges and future trends. *Front. Plant Sci., 8, 471*. doi: 10.3389/fpls.2017.00471.
17. Nair, R., (2016). *Effects of Nanoparticles on Plant Growth and Development*. doi: 10.1007/978–3–319–42154–4_5.
18. Maynard, A. D., et al., (2006). Nanotechnology: The next big thing, or much ado about nothing? *Ann. Occup. Hyg., 51*, 1–12.
19. Nel, A., Xia, T., Madler, L., & Li, N., (2006). Toxic potential of materials at the nanolevel. *Science, 311*, 622–627.
20. Dubchak, S, Ogar, A., Mietelski, J. W., & Turnau, K., (2010). Influence of silver and titanium nanoparticles on arbuscular mycorrhiza colonization and accumulation of radiocaesium in Helianthus annuus. *Span. J. Agric. Res., 8*, S103–S108.
21. Saxena, R., Rajesh, S. T., & Manish, K., (2016). Exploring nanobiotechnology to mitigate abiotic stress in crop plants. *J. Pharm. Sci. and Res., 8*(9), 974–980.
22. Sharma, V. K., Filip, J., Zboril, R., & Varma, R. S., (2015). Natural inorganic nanoparticles—formation, fate, and toxicity in the environment. *Chem. Soc. Rev., 44*(23), 8410–8423.
23. Guo, H., & Barnard, A. S., (2013). Naturally occurring iron oxide nanoparticles: Morphology, surface chemistry and environmental stability. *J. Mater Chem. A., 1*(1), 27–42.
24. Bharde, A., Rautaray, D., Bansal, V., Ahmad, A., Sarkar, I., Yusuf, S. M., et al., (2006). Extracellular biosynthesis of magnetite using fungi. *Small, 2*(1), 135–141.
25. Makarov, V. V., Makarova, S. S., Love, A. J., Sinitsyna, O. V., Dudnik, A. O., Yaminsky, I. V., et al., (2014). Biosynthesis of stable iron oxide nanoparticles in aqueous extracts of Hordeum vulgare and Rumex acetosa plants. *Langmuir, 30*(20), 5982–5988.

26. Wang, L. J., Guo, Z. M., Li, T. J., & Li, M., (2001). The nano structure SiO2 in the plants. *Chin. Sci. Bull., 46*, 625–631.
27. Palmqvist, M. N. G., Seisenbaeva, G. A., Svedlindh, P., & Kessle, G. V., (2017). Maghemite nanoparticles acts as nanozymes improving growth and abiotic stress tolerance in *Brassica napus*. *Nanoscale Research Letters, 12*, 631.
28. Rastogi, A., Zivcak, M, Sytar, O., Kalaji, M. H., He, X., Mbarki, S., & Brestic, M., (2017). Impact of metal and metal oxide nanoparticles on plant: A critical review. *Front. Chem.*, *5*, 78. doi: 10.3389/fchem.2017.00078. eCollection 2017.
29. Rastogi, A., Zivcak, M, Sytar, O., Kalaji, M. H., He, X., Mbarki, S., & Brestic, M., (2017). *Impact of Metal and Metal Oxide Nanoparticles on Plant: A Critical Review*. Front. Chem.|.
30. Maurer-Jones, M. A., Gunsolus, I. L., Murphy, C. J., & Haynes, C. L., (2013). Toxicity of engineered nanoparticles in the environment. *Anal. Chem., 85*, 3036–3049. doi: 10.1021/ac303636s.
31. Barrios, A. C., Rico, C. M., Trujillo-Reyes, J., Medina-Velo, I. A., Peralta-Videa, J. R., & Gardea-Torresdey, J. L., (2016). Effects of uncoated and citric acid coated cerium oxide nanoparticles, bulk cerium oxide, cerium acetate, and citric acid on tomato plants. *Sci. Total Environ., 563&564*, 956–964. doi: 10.1016/j.scitotenv.2015.11.143.
32. Fraceto, L. F., Grillo, R., De Medeiros, G. A., Scognamiglio, V., Rea, G., & Bartolucci, C., (2016). Nanotechnology in agriculture: Which innovation potential does it have? *Front. Environ. Sci., 4*, 20. doi: 10.3389/fenvs.2016.00020.
33. Wang, P., Lombi, E., Zhao, F. J., & Kopittke, P. M., (2016). Nanotechnology: A new opportunity in plant sciences. *Trends Plant Sci., 21*, 699–712. doi: 10.1016/j.tplants.2016.04.005.
34. Joner, E. J., Hartnik, T., & Amundsen, C. E., (2008). Norwegian pollution control authority report no. TA 2304/2007. Bioforsk. *Environmental Fate and Ecotoxicity of Engineered Nanoparticles*, 1–64.
35. Kashyap, P. L., Xiang, X., & Heiden, P., (2015). Chitosan nanoparticle based delivery systems for sustainable agriculture. *Int. J. Biol. Macromol., 77*, 36–51.
36. Dayan, F. E., & Zaccaro, M. L. M., (2012). Chlorophyll fluorescence as a marker for herbicide mechanisms of action. *Pest. Biochem. Physiol., 102*, 189–197.
37. Maruyama, C. R., Guilger, M., Pascoli, M., Bileshy-José, N., Abhilash, P. C., Fraceto, L. F., et al., (2016). Nanoparticles based on chitosan as carriers for the combined herbicides imazapic and imazapyr. *Sci. Rep., 6,* 19768. doi: 10.1038/srep19768.
38. Kharissova, O. V., Dias, H. V. R., Kharisov, B. I., Perez, B. O., & Perez, V. M. J., (2013). The greener synthesis of nanoparticles. *Trends Biotechnol., 31*, 240–248. doi: 10.1016/j.tibtech.2013.01.003.
39. Biswas, P., & Wu, C. Y., (2005). Critical review, nanoparticles and the environment. *J. Air Waste Manag. Assoc., 55*, 708–746.
40. Sanchez-Dominguez, M., Boutonnt, M., & Solans, C., (2009). A novel approach to metal and metal oxide nanoparticle synthesis: The oil-in-water microemulsion reaction method. *J. Nanopart. Res., 11*, 1828–1829.
41. Sintubin, L., De Windt, W., Dick, J., Mast, J., Van Der Ha, D., et al., (2009). Lactic acid bacteria as reducing and capping agent for the fast and efficient production of silver nanoparticles. *Appl. Microbiol. Biotechnol., 84*, 741–749.

42. Huang, J., Li, Q., Sun, D., Lu, Y., Su, Y., et al., (2007). Biosynthesis of silver and gold nanoparticles by novel sundried *Cinnamomum camphora* leaf. *Nanotechnology, 18*, 11–15.
43. Siavash, I., (2011). Green synthesis of metal nanoparticles using plants. *Green Chem., 13*, 2638.
44. Shankar, S. S., Rai, A., Ahmad, A., & Sastry, M., (2004). Rapid synthesis of Au, Ag, and bimetallic Au core-Ag shell nanoparticles using Neem (*Azadirachta indica*) leaf broth. *J. Colloid Interface Sci., 275*, 496–502.
45. Chandran, S. P., Chaudhary, M., Pasricha, R., Ahmad, A., & Sastry, M., (2006). *Biotechnol. Prog., 22*, 577–583.
46. Ankamwar, M. C., & Mural, S., (2005). *Synth. React. Inorg. Met.-Org. Nano-Met. Chem., 35*, 19–26.
47. Badri, N., & Sakthivel, N., (2008). *Mater. Lett., 62*, 4588–4590.
48. Hajirostamlo, B., Mirsaeedghazi, N., Arefnia, M., Shariati, M. A., & Fard, E. A., (2015). The role of research and development in agriculture and its dependent concepts in agriculture (Short Review). *Asian J. Appl. Sci. Eng., 4*, doi: 10.1016/j.cocis.2008.01.005.
49. Das, S., Wolfson, B. P., Tetard, L., Tharkur, J., Bazata, J., & Santra, S., (2015). Effect of N-acetyl cysteine coated CdS: Mn/ZnS quantum dots on seed germination and seedling growth of snow pea (*Pisum sativum*, L.): Imaging and spectroscopic studies. *Environ. Sci., 2,* 203–212. doi: 10.1039/c4en00198b.
50. Perez-de-Luque, A., & Hermosin, M. C., (2013). Nanotechnology and its use in agriculture. In: Bagchi, D., et al., (eds.), *Bio-Nanotechnology: A Revolution in Food, Bomedical and Health Sciences* (pp. 299–405). Wiley-Blackwell, West Sussex, UK.
51. Green, J. M., & Beestman, G. B., (2007). Recently patented and commercialized formulation and adjuvant technology. *Crop Protection, 26*, 320–327.
52. Bhattacharyya, A., Duraisamy, P., Govindarajan, M., Buhroo, A. A., & Prasad, R., (2016). Nano-biofungicides: Emerging trend in insect pest control. In: Prasad, R., (ed.), *Advances and Applications Through Fungal Nanobiotechnology* (pp. 307–319). Cham: Springer International Publishing. doi: 10.1007/978-3-319-42990-8_15.
53. Peyman, A., Masoud, T., Mehrdad, Z., Ivana, T., & Daniel, S., (2015). Effect of SiO_2 nanoparticles on drought resistance in hawthorn seedlings. *Leśne Prace Badawcze / Forest Research Papers, 76*(4), 350–359.
54. Mohamed, A. K. S., et al., (2017). Interactive effect of salinity and silver nanoparticles on photosynthetic and biochemical parameters of wheat. *Arch. Agron. Soil Sci.,* 1–12.
55. Panyuta, O., et al., (2016). The effect of pre-sowing seed treatment with metal nanoparticles on the formation of the defensive reaction of wheat seedlings infected with the eyespot causal agent. *Nanoscale Res. Lett., 11*, 1–5.
56. Taran, N., et al., (2017). Effect of zinc and copper nanoparticles on drought resistance of wheat seedlings. *Nanoscale Res. Lett., 12*, 60.
57. Zheng, L., Hong, F., Lu, S., & Liu, C., (2005). Effect of nano-TiO2 on strength of naturally aged seeds and growth of spinach. *Biol. Trace Elem. Res., 104*, 83–91.
58. Wuttipong, M., Ajit, K. S., Santi, M., & Piyada, T., (2017). Nanopriming technology for enhancing germination and starch metabolism of aged rice seeds using phytosynthesized silver nanoparticles. *Scientific Reports, 7*, 8263. doi: 10.1038/s41598-017-08669-5.

59. Laware, S. L., & Shilpa, R., (2014). Effect of titanium dioxide nanoparticles on hydrolytic and antioxidant enzymes during seed germination in onion. *Int. J. Curr. Microbiol. App. Sci, 3*(7), 749–760.
60. Lu, C. M., Zhang, C. Y., Wen, J. Q., Wu, G. R., & Tao, M. X., (2002). Research of the effect of nanometer materials on germination and growth enhancement of Glycine max and its mechanism. *Soybean Science, 21*, 168–172.
61. Azencott, H. R., et al., (2007). Influence of the cell wall on intracellular delivery to algal cells by electroporation and sonication. *Ultrasound Med. Biol., 33*, 1805–1817.
62. Thornton, P. K., (2010). Livestock production: Recent trends, future prospects. *Philos. Trans R. Soc. Lond. B. Biol. Sci., 365*, 2853–2867.
63. Francis, J. C., Natalie, S. G., Gozde, S. D., Juliana, L. M., & Markita, P. L., (2018). Nanoparticle-mediated delivery towards advancing plant genetic engineering. *Trends in Biotechnology, TIBTEC, 1636*, 16.
64. Rai, M., et al., (2015). Nanoparticles-based delivery systems in plant genetic transformation. In: Rai, M., Ribeiro, C., Mattoso, L., & Duran, N., (eds.), *Nanotechnologies in Food and Agriculture.* Springer, Cham.
65. Rafasnjani, M. S. O., Kiran, U., Ali, A., & Abdin, M. Z., (2016). Transformation efficiency of calcium phosphate nanoparticles for genetic transformation of *Cichorium intybus,* L. *Indian J. Biotechnology, 15*, 145–152.
66. Mou, C. Y., et al., (2013). A simple plant gene delivery system using mesoporous silica nanoparticles as a carrier. *Journal of Materials Chemistry, B., 1*, 5279.
67. Joyner, J. R., & Kumar, D. V., (2015). *Nanosensors and Their Applications in Food Analysis: A Review*, *1*(4), 80–90.
68. Srivastava, A. A., & Surjit, K., (2018). Nanosensors and nanobiosensors in food and agriculture. *Environmental Chemistry Letters, 16*(1), 161.
69. Singh, R. P., (2017). Application of nanomaterials toward development of nanobiosensors and their utility in agriculture. In: *Book: Nanotechnology.* doi: 10.1007/978–981–10–4573–8_14.
70. Saxena, R., Manish, K., & Rajesh, S. T., (2017). Nanobiotechnology: A new paradigm for crop production and sustainable agriculture. *Research Journal of Pharmaceutical, Biological and Chemical Sciences, 8*(4), 823–832.
71. Enisa, O. M., & Mirjana, M., (2016). Nanosensors applications in agriculture and food industry. *Bulletin of the Chemists and Technologists of Bosnia and Herzegovina, 47*, 59–70.
72. Sharma, R., Ragavan, K. V., Thakur, M. S., & Raghavaro, K. S. M. S., (2015). Recent advances in nanoparticle based aptasensors for food contaminants. *Biosensors and Bioelectronics, 74*, 612–627.
73. FAO/WHO (Food and Agriculture Organization of the United Nations/World Health Organization): FAO/WHO Expert meeting on the application of nanotechnologies in the food and agriculture sectors: Potential food safety implications. Rome: Meeting Report, (2010).
74. European Commission (2015). Projects on smart agri food systems. http://ec.europa.eu/newsroom/horizon2020/document.cfm?d oc_id=18158 (accessed on 4/12/2016).

Index

A

Abiotic stress, 214, 219, 223
Acetone, 128
Active drug
  delivery manner, 62
  targeting, 63
Adenosine triphosphate (ATP), 11
Agriculture, 28, 71–73, 78, 81, 82, 210, 212, 213, 215, 217–219, 221–223
  agricultural
    applications, 72
    cost, 222
    environmental systems, 78
    horticultural crops, 202
    management, 218, 221
    practices, 214, 221
    sector, 72, 213, 218, 222
    sustainability, 210, 213, 220
  sustainability, 212
Agrobacterium-mediated gene delivery system, 220
Agrochemical, 210, 213, 215, 217, 218, 221, 222
  wastage, 222
Agronomic
  applications, 210
  traits, 220
AIDS-related Kaposi's sarcoma, 39
Algae, 118, 143
Alginate, 30, 51
Alginic acid, 189
Allografts, 7
Aloe vera, 25, 217
Alternaria alternata, 25
Alternating magnetic field (AMF), 49
Alveolar
  epithelial cells, 99
  membrane, 99
  region, 97
Amines, 11, 49
Amino acid, 27, 148, 149, 152, 154, 216
Amplification, 10, 185, 221
Anthocyanins, 76
Anti-bacterial, 14, 28
  agents, 131
  effect, 131
Antibiotics, 126, 127, 131, 132, 222
Antibody, 5, 10, 46, 49, 52, 63, 64, 68
  fluorescent dye targeting agents, 5
Anticancer, 51, 57, 58, 68, 69
  drugs, 50
  medicine, 58
Antigen-antibody interaction, 49
Antigenicity, 48
Antigens, 64, 68
Antimicrobial
  activity, 4, 10, 127, 128, 206
  agent, 10, 35, 100, 126, 127
  compound, 132
  drugs, 131, 132
  effect, 115, 126, 132
  efficiency, 125, 131
  property, 115, 127, 131, 205
  substances, 127, 131
Antioxidant
  defense system, 77
  enzymes, 80
Antireflecting glasses, 13
Apoptosis, 40, 69, 88, 101, 102
Aptamers, 5, 6, 63
Aqueous
  nonaqueous solution, 128
  solution, 128
Arabidopsis, 4, 221
Arsenic remediation, 203
Asbestos-like pathogenic behavior, 100
Ascorbic acid, 116, 118
Aspergillus clavatus, 25
Atherosclerosis, 39
Atomic level, 138
Autoimmunity, 39
Azadirachta indica, 25, 217

B

Bacillus
cereus, 25
licheniformis, 118
Bacteria, 3, 4, 9, 23–26, 75, 115, 118, 126, 132, 143, 200, 205
Bacterial
cell, 10
infections, 201
Bactericidal property, 115
Basic formal ontology (BFO), 144
Bioaccessibility, 64
Bioaccumulation, 25, 82
Bioactive, 8, 127, 184, 186, 191
bone, 8
compounds, 127
Bioanalytical applications, 202
Biochemical
changes, 77, 81
pathways, 184
Biocompatibility, 48–50, 65, 186, 188
Biocompatible
segments, 192
vesicles, 50
Biodegradable, 14, 15, 50, 61, 62, 65, 185, 188, 193, 194, 215
Biodeterioration, 188
Biofertilizer, 71, 74
Biofilters, 205
Biogenic synthesis, 24
Bioindicators, 82
Bioinformatics, 136, 138, 156
Biological
agents, 8, 118
barrier, 88
effects, 96
molecules, 2–4, 7
pathways, 140
reagent, 119
sources, 24
substitutes, 7
systems, 2, 3, 23, 36, 37, 40, 97, 135, 212
nanoparticles (NPS), 24
Bioluminescent methods, 121
Biomacromolecular components, 5, 10
Biomarkers, 3, 142, 157
Biomass, 13, 26, 74, 81, 82
Biomaterials, 7, 8, 58
Biomedical
applications, 5, 37, 49, 99, 191, 192
clinical applications, 35
equipments, 184
implant, 189
pharmaceutical, 185
physiochemical fields, 125
products, 186
properties, 125
Biomedicine, 23, 141, 191, 194
Biomolecules, 2–4, 9, 10, 24, 29, 46, 64, 145, 202, 216, 220
Bionanoginomics, 137, 157
Biopersistence, 97
Bioreduction, 128, 129, 216
Bio-reduction, 118
Bioremediation, 199, 200, 205
Biosensing, 9, 218
Biosensors, 4, 9, 10, 29, 58, 222
Biosynthesis, 23, 25, 74, 115, 116, 119, 131, 216
Biotechnology, 1–3, 73, 128, 136, 138, 183, 184, 193, 194, 212, 213
Bisphenol A (2,2-bis(4-hydroxyphenyl) propane(BPA), 204
Blood
brain barrier (BBB), 15, 59, 97
circulation, 62
mononuclear cells, 46
solubility, 37
stream, 12, 45, 99
vessels, 38, 190
Boiling point, 51, 171
Bone
injuries, 8
marrow cells, 121
Bottom-up
approaches, 117
system approach, 215
Broad-spectrum
antibacterial nature, 132
antibiotics, 131
Broglie wavelength, 113
Buckminsterfullerene, 201

C

Caenorhabditis elegans, 90
Calcium phosphate, 8
  nanoparticle (CaPN), 220
Camellia sinensis, 119
Cananolab, 141, 157
Cancer, 5, 6, 11, 36, 39, 40, 45–47, 49–52, 57–60, 62, 64, 65, 67, 68, 97, 126, 130, 135–137, 139–141, 144, 146, 156, 157, 201, 203
  cells, 45, 64, 65, 68
  management, 58
  nanotechnology, 68
  progression, 135
  therapy, 57, 65, 68
  treatment, 5, 40, 60, 65
Candida glabrata, 26
Capping agent, 25, 112, 116
Carbides, 201
Carbohydrate concentration, 74
Carbon
  nanoparticles, 201
  nanotubes (CNTs), 4, 10, 40, 72, 88, 89, 96, 99, 142, 143, 145, 201, 206, 207, 216, 217
Carboxymethyl chitosan, 51
Carcinogenic condition, 135
Cardio-respiratory endpoints, 97
Cardiospermum helicacabum, 25
Catalysis, 30
Catalytic properties, 113
Cationic
  liposomes, 39
  silver targets, 10
Cell
  assays, 10
  culture, 120, 188
    assays, 120
  death
    altered gene expression, 40
    gene therapy, 40
  division, 121
  elongation, 74
  membrane, 24, 63, 69, 81, 87, 88, 98
  therapy, 7
Cellular growth, 60
Cellulose, 4, 51, 183, 184, 187
  polymers, 189
Ceramic nanoparticles, 201
Chemical
  abstracts service registry number (CAS RN), 101, 102
  agents, 202
  approaches, 24, 116, 118
  composition, 96
  compound, 116
  methods, 118, 128, 202, 216
  parameters, 96
  structure, 143, 163
  vapor deposition (CVD), 169, 179
Chemotherapeutic medicines, 68
Chemotherapy drugs, 137, 144
Chimeric fusions, 145, 157
Chitosan, 51, 189, 215
Chlorella vulgaris, 119
Chlorophyll, 74–76, 81
Chloroplastic mRNA, 74
Cholesterol, 50
Chondrocytes, 8, 9
Chromatin, 150
Chromium, 82
Chromosomal
  aberration, 121
    test, 88
  rearrangements, 145, 146
  translocation, 146, 157
Chromosome, 121, 146, 147
Cichorium intybus, 220
Circulating tumor cells (CTCs), 45, 52
Cisplatin treatment, 40
Citrate, 116, 118, 194
Cladophora prolifera, 119
Clot removal equipments, 184
Colloids, 111, 113, 114, 189
Colorimetric assay, 121
Combinational therapy, 60
Computational chemistry, 138, 157
Conformational motion, 178
Conventional
  method, 4, 186, 218
  techniques, 79
Copper, 73, 79, 82, 89, 116, 126, 143, 170, 202, 204, 215, 219

Core-shell morphology, 202
Coriandrum sativum, 217
Corneal layer, 98
Corynebacterium strain SH09, 25
Cosmetics, 4, 28, 30, 88, 90, 93, 96, 115, 126, 189
Crataegus, 219
Critical assessment of protein structure prediction (CASP), 147
Crop
  cultivation, 222
  production, 210, 213, 214, 217
  protection, 212, 217, 219, 222
Crude protein, 74
Cryptococcus humicola, 26
Crystal
  perfection, 113
  structure, 120
Crystalline
  macromolecular organic compounds, 163
  regions, 163
  structure, 97
Crystallite size, 120
Crystallography, 145
Cultivation conditions, 79
Culture supernatant, 118
Cuttingedge protein structure modeling, 147
Cyanobacteria, 119
Cymbopogon flexuosus, 217
Cysteine, 151–153
Cytokine production, 87, 88
Cytoplasm, 39, 99, 121
Cytoplasmic
  delivery, 39
  drug delivery, 39
cytosine, 150, 155
  CpG (5' cytosine phosphate-Guanine-3'), 150–153, 155
  islands(CGIs), 150
Cytotoxic properties, 68
Cytotoxicity, 28, 40, 71, 100, 102, 112, 120, 141

D

Dairy industry, 13
Damaged leaf area percentage, 79, 83
Data analysis, 135, 140, 143
Database, 94, 138, 140–143, 157
Degradability, 186
Dendrimers, 37, 60, 62, 68, 116, 142
Dendritic cells, 36
Deoxyribonucleic acid, 200, 207
Dermal absorption, 97
Detrimental effects, 62, 78
Dextran, 51, 52
D-glucuronic acid, 51
Diagnosis, 1, 3, 5, 6, 11, 29, 36, 40, 45–48, 59, 68, 95, 156
Diagnostic devices, 40, 72
Dielectric relaxation, 178
Diels-Birch
  adduct unit, 191
  apportion temperature, 191
  response temperature, 191
Diffusion, 64, 113, 176–178, 186
  pump, 176–178
Digestive tract, 89
Disaccharide unit, 52
Disease
  causing microbes, 127
  management, 213
    plants, 218
Dissolution, 186
DNA, 2, 4, 10–12, 24, 29, 35, 69, 88, 98, 121, 138, 146, 148, 150–153, 155, 157, 188, 200, 220, 222
  computer circuits, 2
  structure, 35
Doxorubicin, 39, 50
Drug, 1, 2, 4, 7, 11, 28, 30, 35–41, 48–52, 57–65, 69, 90, 126, 131, 136, 139, 140, 144, 146, 147, 150, 157, 183–186, 188, 189, 192–194, 201–203, 220
  carrier's devices, 52, 184
  concentration, 64, 186
  delivery, 1, 7, 11, 30, 35–41, 50, 58–63, 65, 90, 126, 136, 144, 183–185, 188, 189, 192–194, 201, 202
  loading capacity, 61
  packaging, 195
  platform, 48
  products, 7
  targeting, 36, 41, 60, 64, 193
    strategies, 62

transport
mechanisms, 64
process, 64
transportation, 186
Dynamic light scattering (DLS), 26, 28

E

Echogenicity, 51
Eco-friendly, 24, 116, 118, 125, 131, 199, 206, 216, 217, 222
approach, 24, 216, 222
Ecological hazard estimation, 143
Ecosystem, 82, 90, 102
Ecotoxicity, 142, 157
Electric field, 37, 178, 179
Electrical
assays, 10
property, 113
technique, 173
voltage, 8
Electromagnetic energy, 49
Electron
affinity, 201
beam, 28, 170
bombardment, 169
donor functional groups, 11
gun, 120, 170
microscopy, 120, 145
Electronic
biomolecule detectors, 10
circuit, 173
devices, 16, 87
impulses, 11
products, 130
sensors, 10
structure, 2, 113
Electroporation, 220
Electrospinning, 8
Electrospun nanofibers, 7
Ellipsometric method, 172
Emulsifiers, 203
ENanoMapper ontology, 143, 157
Encapsulation, 37, 38, 61, 65, 144, 193, 218
Encephalopathy, 89
Endocytic delivery, 39
Endocytosis, 63, 64
Endo-lysosomal compartment, 39
Endothelial cells, 36, 39
Energetic
biomass employment, 13
particles, 166, 167
Energy
dispersive x-ray analysis (EDX), 26, 28, 81
generation, 13
utilization, 13
sector, 130
transfer measurement, 37
Engineered nanomaterials (ENMs), 81, 140, 141, 143, 200
Enhanced permeation and retention (EPR), 60, 63
Environmental
cleaning, 203
conditions, 4, 71, 74
issues, 14, 217
management, 199, 206
stress, 9
Environment-friendly
approach, 25, 128
nanomaterials, 143
Enzyme
activities, 30, 77, 82, 121
coproporphyrinogen oxidase, 74
system, 214
specific drug, 51
Epidermal
cell walls, 81
growth factor receptors, 5
Epithelioid granuloma, 15
Equilibrium state, 166
Escherichia coli, 9, 25, 90
Ethanol, 128
Ethylene glycol, 50
Eukaryote, 200
Evaporation-condensation, 118
Extracellular
enzymes, 25, 26
matrix, 8
synthesis, 26

F

Fabrication, 15, 16, 185, 200
Fat cells, 60

Ferric nanoparticles, 81
Fertilizer development, 210
Fertilizers, 13, 75, 83, 126, 213, 215, 217, 218, 222
Fibroblasts, 8, 100
Film
  condenser, 173
  formation, 169, 170
  polymer solution, 170
  preparation, 171, 172
  specimen, 173–175
  substrate, 165
Flocculent, 187
Fluorescein isothiocyanate(FITC), 10
Fluorescence
  emission, 37
  imaging techniques, 6
  properties, 10
  resonance, 37
Fluorescent
  dyes, 10
  in situ hybridization (FISH), 10, 59
Foliar spray, 74
Folkloric application, 127
Food
  additive, 115
  FDA (Food and Drug Administration), 50, 64, 68, 115, 189, 195
  industry, 126, 222
  packaging, 221
Foodborne
  diseases, 4
  pathogens, 9
Fourier transform infrared (FTIR) spectroscopy), 26, 27, 30, 112, 119
Full width at half maximum (FWHM), 120
Fullerenol-lysozyme complex, 145
Fungal mycelial growth, 79
Fungi, 23–26, 75, 118, 119, 126
Fusarium
  acuminatum, 25
  oxysporum, 118, 119
  semitectum, 25
  solani, 25

G

Gastrointestinal (GI) disorders, 15, 97, 99
Gelatin, 114
Gene
  delivery
    system, 220
    vehicles, 220
  therapy, 1, 3, 6, 16, 65, 68
  transfer, 40, 72
  transformation, 212, 220
Generally recognized as safe (GRAS), 115, 122
Genes facilitates, 144
Genetic
  delivery systems, 210
  engineering, 64, 210, 220
  transformation, 213, 220
Genome, 6, 40, 140, 150
Genomic
  data, 157
  medicine, 140, 157
Genotoxic, 121
Genotoxicity, 28, 87–90, 112, 120, 121
Germ cell, 90
Germination, 73
  index, 75, 80
  of rice seeds, 77
  percentage, 75, 79
  rate, 80, 81
Glaucoma, 201
Globally harmonized system (GHS), 121, 122
Glow discharge polymerization, 169
Glutathione (GSH), 80, 83
Glycol chitosan, 51
Glycyrrhiza glabra, 4
Gold nanoparticles (AuNPs), 5, 10, 16, 37, 47, 48, 52, 88, 100, 113, 115, 116, 125, 129, 131, 132, 216
Granulomas, 89, 100
Green nanoscience, 101
Growth and development, 73–75, 79, 163, 214

H

Health surveillance systems, 101
Hematopoietic cells, 146
Hemolysis assay, 28
Heparin, 51, 52
Herbagreen nanoparticle, 73, 83

Herbicides, 213, 215, 217–219, 222
Heterochromatic regions, 150
Hydrobenzamide, 116
High-frequency dielectric constant, 173
Hind Hi Vacuum Pumping System, 172
Homology modeling, 147, 148
Human
immunodeficiency virus (HIV), 6, 67, 131, 132
HIV-1 virus, 6
pathogens, 127, 132
Hyaluronic acid, 51
5-β-cholanic acid, 51
Hyaluronidase-1, 51
Hydrocarbon mixtures, 13
Hydrocolloids, 189
Hydrophilic, 50, 51, 62, 183, 188
Hydrophilicity, 97
Hydrophobic
behavior, 186
cargo, 65
chemodrugs, 50
hydrophilic core, 62
molecule, 51
Hydrophobicity, 152
Hydroponic
conditions, 81
plant growth, 81
Hydroxyethyl, 189
Hydroxyls, 11
Hydroxypropyl
cellulose, 189
methylcellulose, 189
Hyperthermia, 4, 49, 65

I

Imidazoles, 11
Immobilization, 10, 30, 179, 221
Immune
response, 61
system, 52
Immunogenic reactions, 60
Immunogenicity, 48, 58, 64
Immunosuppressive agents, 131
Immunotoxicity, 40, 141
Indoles, 11
Infant's leukemia, 146, 157
Inflammation, 15, 39, 88, 89, 98, 100, 102
Inflammatory
bowel disease, 15
effects, 87
Inhalation, 88, 97, 121, 122, 130
Injectable biopolymers, 7
Interferometric method, 172
Intravenous
catheters organ transplantation, 131
delivery, 37, 38
Investigation study assay (ISA), 142
TAB open-source framework, 142
tab-delimited format (ISA-TAB-Nano), 142
TAB-nano, 142, 157
file arrangement, 142
Ion
bombardment energy, 168
exchange, 186
Ionic
bombardment, 177
ligands, 48
Ionization, 166, 167, 177
Iron, 29, 37, 47, 49, 52, 62, 63, 74, 77, 79, 80, 82, 126, 143, 203–206, 214, 215
oxide, 29, 47, 52, 63, 143, 214
nanocrystals, 37
nanoparticles (IONPs), 29, 49, 63, 214
sulfate, 74
sulfide NPs, 203
Irradiation reduction, 128
Island growth mode/volmer-weber mode, 165
Isothermal
immersion technique, 171
technique, 172
I-TASSER software, 148, 157

K

Keratinocytes, 8, 40
Kinetic property, 30
Kinetics, 58

L

Lactide-co-glycolide polymers, 187
Laser
ablation, 118, 169

Frank-Vander Merwe mode, 165
plus-island growth mode or stranski-krastanov mode, 165
weapons, 12
Lazy Structure-Activity Relationships, 143
Lemongrass, 116, 217
Leukemogenesis, 146, 147, 150, 158
Life threatening diseases, 130
Ligand molecules, 139
Light sectioning method, 172
Lipid
bilayer membrane, 50
materials, 50
membrane, 4, 10
nanoformulation, 50
nanoparticles, 201, 203
peroxidation, 77, 80, 81
Lipophilic molecules, 203
Liposomal vesicles, 50
Liposomes, 37, 50, 60, 62, 63, 68
Liquid
colloid, 114
contact electrodes, 174, 175
dosage, 186, 194
Listeria monocytogenes, 9
Localized surface plasmon resonance (LSPR), 47, 52
Luciferase enzyme, 121
Lupinus termis, 219
Lymphatic drainage, 63
Lyngbya majuscule, 119
Lysosomes, 39, 49

M

M. grisea, 79
Macrophage, 38, 39, 98
contribution, 39
destruction, 39
Macroscopic level, 185
Mad's cow disease, 89
Magna-porthe grisea, 78
Magnesium oxides, 204
nanoparticles, 83
Magnetic
field, 37, 49, 144, 184
properties, 49
resonance imaging (MRI), 4, 37, 49, 52, 59, 65
Magnetite, 49, 73, 82, 204
Magnetron sputtering, 169
Magnitude, 113, 154, 179
Maize, 71, 73–76, 82
Malignant transformation, 145
Malondialdehyde content, 80
Mammalian cells, 6
Mandatory monitoring schemes, 101
Manganese oxides, 204
Medical
devices, 14, 65, 93, 96, 126
diagnostic devices, 2
grade adhesive for transdermal delivery, 188
imaging, 2, 5, 97
Medicinal
agents, 127
plants, 127, 131
Mercaptopropionic acid, 100
Mesoporous molecular sieving(MCM), 204
Mesoporous silica nanoparticles (MSNs), 48, 52, 72, 221, 223
Mesothelioma, 88
Metabolic alkalosis, 89
Metallic
cations, 5
ions, 5, 10
nonmetallic substrate, 175
surface, 175
Methyl-CpG-binding domain (MBD), 150, 158
Michelson's interferometer, 172
Microbial
activity, 89, 199, 203
systems, 28
Microcarriers, 185
Microemulsion method, 128
Micro-environment, 7, 9
Micrometers, 8, 164, 173
Microminiaturization, 163
Microns, 6
Micronuclei, 121
Micronutrients, 74, 215
Microorganism, 127, 129

Microscope, 94, 155, 173
Microscopic
nanoscale, 120
observations, 80
Microscopy techniques, 26
Microstructural characteristics, 169
Microwave-assisted synthesis, 128
Military applications, 11
Minerals, 114
Mirabilis jalapa, 25
Misfolded region, 149
Mitochondrial activity, 89
Mixed-lineage leukemia (MLL), 146–150, 152, 155, 156
Mobilization, 80
Model 12A-4, 176
Molecular
biology, 1, 12, 64
chains, 51
dynamics (MD), 36, 147, 148, 150, 152–156, 158
level, 4, 58, 60, 135, 140, 145
modeling, 145, 158
structures, 111, 112
Molybdenum boat, 176
Monocharge electrets, 175
Monochromatic light, 28
Monoclonal antibodies, 5
Monomer gas molecules, 169
Mononuclear phagocyte, 36
system (MPS), 63
Montmorillonite (MMT), 114, 122
Morbidity, 99
Morphology disorder, 81
Mortality, 97, 99
Magnetoresistive Random Access Memory (MRAM), 12
Multidisciplinary application, 72
Multi-drug resistant
bacterial strains, 28
microbial strains, 131
Multifunctional approaches, 58
Multivalency, 58
Multi-walled carbon nanotubes (MWCNTs), 77, 201, 207
Murraya koenigii, 25
MVD software, 151, 158
Myeloid progenitor cells, 146

N

N-(hydroxymethyl) acrylamide, 191
N-acetyl-D-glucosamine, 51
Nanoanalytical contrast reagents, 57
Nanoand microparticles, 188
Nanobarcodes, 222
Nanobioinformatics, 137, 145, 158
Nanobiosensors, 212, 213, 221–223
exhibit high sensitivity, 221
Nanobiotechnology 1–5, 11, 15, 16, 50, 57, 58, 68, 136, 137, 158, 210, 212–216, 218–220
crop protection, 218
future prospectus of nanobiotechnology, 11
medical applications, 5
diagnosis, 5
drug delivery, 7
gene therapy, 6
pathogen detection, 9
tissue engineering (TE), 7
Nanobots, 12
Nano-builder, 2
Nanocapsular, 202
particle, 202
type, 202
Nanocapsules, 62, 98
Nanocarrier, 37–39, 50, 57, 61–63, 68, 69, 144, 192, 215
mediated macrophage, 39
Nanoceramic, 201
Nanoceuticals, 13
Nano-chelate zinc, 75
Nanochemical moieties, 72
Nanoconjugate, 64
Nanocrystals, 49, 100
Nano-data discovery and extraction, 142
Nano-devices, 6
Nanodrug, 57, 59, 60, 62, 64, 65, 68, 69, 130, 139
Nanoecotoxicity, 78
Nanoelectromechanical systems, 65
Nanoemulsions, 50
NanoE-TOX, 142, 158
database, 143
Nanofabricated devices, 212

Nanofactories, 12, 126
Nanofertilizers, 213, 223
Nanofibers, 36
Nanofibrous units, 191
Nanoherbicides, 213, 222
NanoHub, 13
Nanoinformatics, 135–140, 144, 150, 156–158
  analysis, 139
  databases, 136
Nanolayered corrosion inhibitors, 12
Nano-lazar, 143, 158
Nanomachines, 36, 112
Nanomaterials, 1–4, 9, 11, 12, 14, 23, 24, 28, 35, 36, 46–48, 78, 81, 88, 93–96, 98, 101, 111–114, 119, 128, 136–138, 140–145, 157, 199–201, 203–206, 214, 215, 217, 222
  electron microscopy, 120
  uv-visible spectroscopy, 119
  XRD, 120
Nanomedicine, 3, 36, 40, 41, 46, 47, 57–60, 62, 64, 66–69, 132, 140, 144, 148, 158
  clinical applications, 57
Theranostic nanomedicines, 47
  inorganic, 47
    gold nanoparticles, 47
    iron-oxide nanoparticles, 49
    mesoporous silica, 48
  organic nanoparticles, 50
    lipid-based nanoparticles, 50
    polymer-based nanoparticles, 50
Nanomembranes, 13
Nanometer, 2, 6, 94, 112, 136, 164, 200
Nanometric thin films, 12
Nano-mineral supplementation, 77
Nano-moieties, 77
Nanomolecules, 77
Nanooncology, 58
Nanoparticle, 1, 4, 5, 12, 14, 15, 23–25, 27, 29, 30, 35, 39, 46, 49, 50, 52, 57, 61, 63, 67, 69, 71–76, 78–83, 94–100, 102, 111, 113, 115, 116, 118–120, 122, 125, 127–130, 132, 136, 137, 169, 179, 188, 195, 200–203, 207, 210, 212–216, 219, 220, 127–129
  application, 71
  characterization, 111, 119, 120
  imaging, 46
  material, 113
  mediated bonding, 99
  ontology (NPO), 143, 144
  products, 71
  remediation, 203
  synthesis
    bacteria, 25
    fungi, 25
    plant extracts, 25
Nanoparticular materials, 13
Nanopesticides, 213, 215, 222
Nanopharmaceuticals, 57
Nanopore filters, 113
Nanoporous, 6, 13
Nanoprocessing, 222
Nanorobots, 12
Nanoscale, 2, 3, 6, 7, 37, 60, 68, 94, 101, 111, 113, 115, 120, 138, 200, 212, 213
  materials, 2, 36
  microfabrication-based devices, 36
  molecular signals, 7
  properties, 112
Nanoscience, 73, 94, 136, 158
Nanosensors, 13, 57, 72, 212,213, 221, 222
Nano-silicon cells, 13
Nanosilver, 80, 115
Nanosize devices, 16
  gene carriers, 6
Nanosized
  formulation, 74
  zero-valent ion (NZVI), 203, 207
Nanospheres, 62, 98, 202
  particle, 202
Nanostructure, 94, 114
Nanostructured
  fillers, 12
  materials, 99
  particles, 94
Nanosuspensions, 50
Nanosystems, 37, 38, 60, 65
Nanotechnological methods, 65
Nanotechnologies for medical imaging, 36
Nanotechnology based drugs, 59
Nanotools, 13, 37

Nanotoxicity, 96, 97, 102
Nanotoxicological studies, 95
Nanotoxicology, 88, 90, 93, 96, 97
Nanotube, 1, 10, 15, 40, 72, 77, 88, 89, 94, 96, 101, 102, 130, 142, 201, 205, 206, 216, 217
  electronic devices, 10
Nanovehicles, 37, 41
  drug carriers, 37
Nanowires, 65, 94
Nano-ZnO, 78
National Cancer Institute (NCI), 40, 141
  Nanotechnology Characterization Laboratory, 141
  Office of Cancer Nanotechnology Research, 141
Natural
  environment, 199, 206
  nanomaterials, 113
  pharmaceutical biopolymers, 184
  polymer, 51
  products, 127
  routes, 38
  sources, 114, 127
Neovascularization, 51, 59
Newton's rings, 172
Next generation computer processing, 12
N-isopropyl acrylamide, 191
Nitrides, 201
Non-biodegradable pollutant, 15
Nondestructive method, 172
Non-enzymatic antioxidants, 81
Non-invasive
  photothermal therapy, 46
  tumor detection, 68
Nontoxic
  biproducts, 46, 183
  products, 188
Novel molecular materials, 58
Nuclear
  collisions, 168
  magnetic resonance, 145
Nucleic acids, 3, 5, 9, 10, 40, 136, 145
Nucleotide level, 135, 140
Nutrient applications, 202
Nutritional
  requirements fungi, 26
  uptake, 217
  value, 71, 79

O

Office of Cancer Nanotechnology Research (OCNR), 141
Oil industry, 12
Olfactory mucosa, 97
Oncogenic
  fusion proteins (OFPs), 145, 146, 150
  protein structure, 158
Oncology, 68
Ontologies, 136, 138, 140, 141, 143, 158
Ontology web language (OWL), 144
OpenTox framework, 143
Organelles, 4, 6, 87, 88, 98
Organic
  dyes, 47
  hybridization, 48
  pollutants, 204
  solvents, 52
Organization for Economic Cooperation and Development (OECD), 121, 122
Organometallic compounds, 62
Oryza sativa, 77, 78
Oryzae bacteria, 78
Oryzias latipes, 89
Oscillation, 26
Osmosis, 186
Oxidative stress, 39, 40, 77, 81, 89, 98, 101, 102, 214
Oxygen stress, 87, 88

P

Paclitaxel (PTX), 49
Palladium, 126
Parkinson's diseases, 8
Passive
  active targeting methods, 68
  drug delivery manner, 62
  targeting, 62, 63
Pathogen, 1, 4, 9, 10, 16, 78, 217, 218, 222
Pathogenic
  disease, 127
  microorganisms, 128, 132
P-chloroacetophenone oxime, 191

Pectinic acid, 189
Pelargonium graveolens, 216
Penicillium
  brevicompactum, 25
  fellutanum, 25
Penning gauge, 176, 178
Peptides molecules, 222
Perfluoropentane, 51
Peribronchial inflammation, 15
Peroxidase activities, 82
Personalized medicine, 140, 158
Pest management, 221
Pesticides, 13, 75, 126, 213, 215, 217–219, 222
Petri-dishes, 188
Phagocytosis, 63
Phagokinetic studies, 4
Pharmaceutical
  companies, 38
  field, 185
  polymers, 185, 187, 194, 195
Pharmacokinetics, 37, 58
Pharmacologically active components, 127
Phenylalanine, 88
Phosphates, 11, 201
Phosphatidylcholine, 50
Phosphodiesterase, 49
Photobleaching, 47, 65
Photocatalysis, 115, 201, 202
Photocatalytic
  degradation, 204
  reduction, 128
Photochemical, 127, 202
Photodegradation, 201
Photodynamic therapy (PDT), 65, 68, 69
Photolytic polymerization, 170
Photons, 167
Photo-optics, 202
Photoreduction activities, 76
Photosynthesis improvement, 74
Photosynthetic
  efficiency, 76
  material, 76
  stress, 77
Photosystem II, 76
Photothermal, 46, 47
Photovoltaic, 13
PH-sensitive polymers, 48
Physcio-chemical, 158
Physical
  chemical
    compounds, 125
    methods, 127, 216
    properties, 2, 50, 72, 111, 112, 137, 194, 201
  methods, 118, 125, 128, 131
  nature, 98
  vapor deposition, 164, 179
    technique (PVD), 164, 166, 169
Physicochemical
  characteristics, 152
  properties, 29, 112, 132, 214, 215
Phytocompounds, 127, 128, 131, 132
Phytomolecules, 131
Phytonanoparticles, 131
Phytotoxic, 73, 75, 78, 81
Phytotoxicity, 73, 77, 81
Piezoelectric biosensor, 10
Pirani gauge, 176, 177
Plant
  fertilizer, 214
  growth productivity, 217, 219
    regulation, 74
  hormonal activities, 75
  mediated
    NP synthesis, 25
    synthesis, 25
  protection devices, 72
  substances, 8
  system, 220
Plasma
  concentration, 186
  replacement, 187
Pleurotus sajorcaju, 25
Polarization, 178
Pollutant degradation, 205
Poly
  (DL-lactic-coglycolic acid) (PLGA), 7
  (DL-lactide-co-glycolide), 39
  (ε-caprolactone), 192
  (ethylene oxide), 187
  (ethylene-co-vinyl acetate), 7
  chloroprene, 188
  cyclopropyl methacrylamide, 187

hydroxy ethyl methacrylate, 188
isopropyl poly-acrylamide, 187
(methyl methacrylate), 7, 188, 194
co-[poly(methyl methacrylate)-graft-(oligocaprolactone)] urethane systems, 191
Polyamide (nylon), 187
Polycarbonate, 187
Polyesters, 187
Polyethylene, 10, 48, 63, 187, 192, 193, 206
glycol (PEG), 48, 52, 67, 68, 100, 187, 192, 195
oxide, 10
Polyimide, 7
Polyisobutylene pressure-sensitive adhesives, 188
Polymer, 7, 50, 61, 115, 116, 163, 169–172, 174–176, 183–189, 193, 194, 202, 203, 215
degradation, 183
shell, 202
solution, 7, 169, 171
Polymeric
compounds, 7
conjugates, 60
materials, 186, 188, 191, 194
matrices, 8
micelles, 40, 60
nanoconstructs, 36
nanomedicine, 50
nanoparticles, 201, 202
tools, 194
Polymerization, 169–172, 179
electron bombardment, 170
Polymers, 7, 8, 28, 48, 50–52, 61, 62, 163, 169, 175, 178, 179, 183–194, 215
drug packaging, 186
controlled release dosage form, 186
standard dosage, 186
Polypeptides, 145, 146, 158
Polypropylene, 170, 187, 193
Polysaccharides, 51, 61, 216
Polystyrene, 187, 188
Polyurethanes, 187, 191
Polyvinyl
alcohol, 7, 187, 192
chloride, 187, 188, 194
pyrrolidone (PVP), 90, 116, 122, 187
Polyvinylidene chloride, 187
Posttreatment mitosis, 121
Power utilization, 13
Proline contents, 80
Prominentlyidyllic receptors, 64
Protease, 82
Proteins, 3–5, 9, 10, 27, 29, 38, 51, 61, 74, 79, 83, 89, 119, 135, 145–147, 150, 153, 216, 222
biomarker, 6
delivery, 187
misfolding, 98
peptide drugs, 52
protein binding, 10
residues, 154
sequence, 138, 158
structure, 139, 147, 151
Proteolysis, 48
Proteomic data, 139, 158
Pulmonary deposition, 97

Q

Qualitative concepts, 137
Quantitative spectroscopic technique, 27
Quantum
bits, 12
computers, 12, 130, 136
dots (QDs), 6, 10, 11, 16, 37, 38, 40, 59, 63, 65, 68, 94, 102, 216–218
smart drugs for cancer treatment, 11
effects, 2
Quartz-sand, 82
Qubits, 12

R

Radiofrequency therapy, 65, 68
Radius of gyration (ROG), 156
Ramachandran plot values, 149
Raman probe, 59
Receptor-mediated endocytosis, 64
Redox sensitivity, 87, 88
Remediation, 199, 203–205
Remobilization, 76
Renewable
energy, 13
resources, 203

Reproductive stages, 76
Residual solvent, 172
Respiratory enzymes, 24
Retro-transposition, 146
Revolutionalized proposals, 13
Rhizoclonium heiroglyphicum, 119
Rhodosporidium diobovatum, 26
Rice
blast disease, 78
germination ratio, 77
RNA, 6, 69, 150, 158, 188, 203
to RNase degradation, 6
Rotary pump, 176–178
Route of toxicity, 88

S

Saccharomyces cerevisiae, 26
Salinity stress, 219
Scanning electron microscopy (SEM), 26–28, 30, 79, 81, 120, 122
Schizosaccharomyces pombe, 26
Sclerotinia sclerotiorum, 25
Seed germination, 75, 78, 80, 82, 89, 217, 219
Seedling
growth, 89, 217, 219
index, 75
Semiconductor NPs, 201, 202
Sensory axons, 98
Serum instability, 6
Sign of purity, 114
Silica nanoparticle, 48
Silicon dioxide (nSiO), 73
Silicones pacifier, 188
Silver
ions, 11, 80, 118, 206
nanomaterial, 128
nanoparticles (AgNPs), 4, 25, 61, 72, 75, 78–81, 90, 115, 116, 119, 125, 128, 129, 132, 204, 205, 216, 219
sulfadiazine ointment, 30
treated cells, 80
Simulation, 137, 148, 150–154, 156–158
Single cross 704, 76, 83
Single-stranded nucleic acid molecules, 6
Single-walled
carbon nanotubes (SWCNTs), 10, 89, 90, 201, 207
nanotubes, 89
Skin irritation, 101
Smart drugs, 1, 11, 130, 144, 158
Sodium boro-hydride (NaBH), 118
Soil
absorption, 74
application, 75
fertility, 217
harmony, 217
microbes, 214
organisms, 101
Solid
fiber, 8
liquid interface, 174
state laser system, 12
tumors, 39
Solubility, 96
Soluble carbohydrates, 74
Spectroscopy techniques, 26
Spectrum application, 220
Spinal cord injury, 8
Spin-spin relaxation, 37
Spiral tungsten filament, 178
Spirulina subsalsa, 119
Sputter deposition, 166
Sputtering, 166–168
process, 166
SRB assay, 121
Stabilization, 116, 119
Staphylococcus aureus, 25
Starch-based
biodegradable polymers, 188
polymer, 187, 188
Sunburns, 88
Sunscreens, 30, 100
Superhydrophobic nature, 114
Superoxide dismutases activity, 80
Superparamagnetic particles, 113
Superparamagnetism, 113
Surface plasmon resonance (SPR), 26, 30, 112
Sustainable agriculture, 223

T

Tab-delimited (TAB), 142
Tablet formation, 187
Tamarindus indica, 217
Tanning, 127
Target identification, 139

Targeting
method, 63, 64
solid cancers, 39
Taxonomies, 137
Theoretical methods, 145
Theranostic
agents, 46, 51
applications, 48
efficacy, 58
nanomaterials, 46
Theranostics, 45, 46, 51, 65, 68
Therapeutic
action, 46, 147
agents, 36, 40, 46, 49, 50
approach, 46
compounds, 46
devices, 188
effect, 57, 68
index, 58, 69
intervention, 48
method, 7
molecule, 52
treatment, 57, 131
Thermodynamic stability, 150
Thermoelectrets, 178, 179
Thiols, 11, 77
Thrombosis, 39
Tissue
damage, 220
distribution, 58
engineering (TE), 1, 3, 4, 7–9, 16, 35, 58, 183, 184, 188, 195
regeneration, 5, 58, 130
Titanium
dioxide, 30, 73, 76, 80, 143, 204, 206
dioxide NPs, 204
oxides, 204
Top-down approach, 2, 111, 117, 215
Toxic
byproducts, 128
cadmium ions, 40
chemicals, 128
effects, 75, 80, 99
emission, 12
material, 38, 199, 203
Toxicants, 25
Toxicity, 1, 5, 7, 10, 15, 16, 28, 40, 41, 48, 50, 59, 61, 62, 73, 74, 77, 82, 87, 89, 90, 93–96, 98, 99, 101, 102, 112, 115, 120–122, 125, 127, 140–145, 194, 204, 206, 207, 215, 219
acute and sub-acute toxicity, 121
cytotoxicity assay, 120
genotoxicity assay, 121
issues, 40
mechanism, 73
Toxicological data, 142, 143
Toxicology, 96, 137, 143
Transcriptomic data, 158
Transdermal delivery, 188
Transform plants, 220
Transmission electron microscopy (TEM), 26, 27, 30, 81, 112, 119, 120, 122
Transmitting information, 16
Transparent films, 172
Transplantation, 3, 7
Traumatic processes, 8
Treatment therapy, 69
Tribo-mechanical activation, 73
Tri-sodium citrate, 116
Triticum aestivum, 79–81
Tryptophan, 74
Tumor, 4, 6, 37, 45, 46, 48, 49, 51, 52, 57, 60, 62–65, 68, 100, 144
cell, 37, 45, 49, 51, 52, 63–65, 100
destruction, 4
molecular profile, 68

U

Ultrasonic-assisted reduction, 128
Ultrasonography, 59
Ultrasound imaging applications, 51
Ultraviolet
absorber, 115
irradiation, 40
radiation, 169, 170
rays, 11, 170
visible spectroscopy, 26, 112, 119, 120
Ureibacillus thermosphaericus, 25

V

Vaccines, 137

Vacuum, 27, 164, 169–171, 173, 175–178, 204
chamber, 176, 177
coating of polymer films, 176
operation of the unit, 177
vacuum coating unit, 176, 177
deposited electrodes, 175
evaporation, 169, 170
Vascular stents, 184, 185
Vasculatures substances, 68
Vegetable and food preservation, 30
Vegetative stage, 76
Viscosity, 186

W

Wastewater treatment, 126, 204
Water
droplets, 114
fleas, 89
insoluble, 187, 188
biodegradable polymers, 187
polymers, 187
management, 72
molecules, 2, 37
soluble, 187
packaging, 187
polymers, 187
stressed conditions, 71
treatment processes, 206
Wavelength, 27, 28, 37, 129
Web-based
forms, 142
portal, 141
Weighing method, 173
Wheat
germination, 80
plants, 81, 82
Wounds, 8, 30, 88
healing, 8, 30

X

Xanthomonas oryzae, 78
Xenografts, 7
X-rays, 28, 144, 167
based characterization techniques, 26
diffraction patterns, 120
photoemission spectroscopy (XPS), 26, 27, 30

Y

Yeast, 26
Young's equation, 165

Z

Zea mays, 75, 76
Zeolites, 13
Zero
magnetism, 49
valent iron particles (ZVI), 204, 205
ZF-CXXC domain, 151
Zinc, 10, 30, 73–77, 82, 88, 115, 116, 126, 143, 150, 204, 215, 219
nano-chelate foliar, 75
oxide (ZnO), 30, 73, 76, 82, 88, 89, 100, 115, 143, 202, 204–206, 216, 217
nano-colloid, 76
nanoparticles, 115
particle, 76
sulfide, 10, 100, 202
Z-score, 149